AF801628

Crystal Research and Technology

11

Journal of Experimental and Industrial Crystallography

Zeitschrift für experimentelle und technische Kristallographie

Volume 22
1987
Number 11 November

Akademie-Verlag · Berlin

ISSN 0232-1300 Cryst. Res. Technol., Berlin 22 (1987) 11, 1341–1442, K173–K218

Directions to the author

In the journal "Crystal Research and Technology – Kristall und Technik" as a rule the following categories of papers will be printed: original papers, review articles, short notes, and book reviews, **preferentially in English.** Complete information necessary for type-setting and a "Short Title" should be added.

The manuscript should not exceed 10 pages and be type-written on one side only in the format A4 (21 × 29.7 cm) on white ink-proof paper, with a two-lined distance and a left-hand free edge of 3 cm, and submitted in 2 copies. All sheets of paper belonging to the contribution are continuously to be numbered.

The head of the manuscript should be arranged in this manner: initials and name of the author; institution, and title of the paper. Then two abstracts should follow, one in the original language and the second in German, English, French, or Russian, respectively.

Formulae, signs and symbols, letters etc. being wanting on the type-writer, should be distinctly written with penholder. O as zero is to be underlined, the capital O is not. Greek letters should be underlined with red ink.

Paragraphs should be marked by indenting of at least 3 touches.

Concerning the marking of the manuscript should be noted:
headings should be underlined, sub-headings not, and indented by 9 touches. The titles of tables are not to be underlined. Other markings in the text will be done by the editorial staff. They mean:

light-red ∼∼∼∼∼∼∼∼ = small capitals
blue ∼∼∼∼∼∼∼∼ = in italics

light-red | perpendicular straight red line on the left edge = small printing

Tables are to be added on separate sheets of paper at the end of the contribution.

Foot notes should be continuously written on a separate sheet of paper.

Figures (Fig.) – drawings and photographs – should be not more than 8 and submitted (2 copies) in a special envelope: photographs as white high-gloss prints (inscription on a transparent covering sheet) and drawings as distinct sketches. On the back of the submitted figures the reduction scale in tenth (9/10, 8/10, 5/10 e.g.), author's name, short title of the paper, and the number of the figure should be given.

Captions for the figures should be submitted on a separate sheet of paper.

References should be written without numbers in alphabetical order of the names at the end of the contribution. Cited names in the text stand for the references. The references should contain the following items, for example:

for books:	RAST, K.: Physikalisch-chemische Rechnungen, Berlin 1958
for journals:	KRAUSE, H. N., MÜLLER, K.: Cryst. Res. Technol. **16** (1981) 70

Add, please, the short title, about 6 keywords, and the academic grades of the authors.

The authors will be sent two galley proofs of review articles and original papers. Proof reading of short notes and book reviews will be done by the editorial staff.

For each of the papers a total of 50 reprints is available free of charge. Requests for additional maximum 200 reprints (against payment) should be given together with the submitted manuscript.

11

Crystal Research and Technology

Founded 1966 as
Kristall und Technik

Journal of Experimental and Industrial Crystallography

ESTABLISHED BY
W. KLEBER and H. NEELS

EDITED BY
H. NEELS, Leipzig

EDITOR-IN-CHIEF
H. NEELS, Leipzig

VOLUME 22 · 1987 · NUMBER 11

ADVISORY BOARD
L. N. ALEKSANDROV, Novosibirsk
Č. BARTA, Prague; F. BERTAUT, Grenoble
G. BLIZNAKOV, Sofia; J. BOHM, Berlin
O. BRÜMMER, Halle
V. A. FRANK-KAMENETZKIJ, Leningrad
S. HAUSSÜHL, Köln
J. HEYDENREICH, Halle
R. KAISCHEW, Sofia; R. KERN, Marseille
Y. KOMURA, Hiroshima; I. KOSTOV, Sofia
P. KRISHNA, Varanasi
K. ŁUKASZEWICZ, Wrocław; K. MEYER, Berlin
P. PAUFLER, Leipzig; I. S. REZ, Moscow
W. SCHATT, Dresden, N. N. ŠEFTAL' †, Moscow
D. SCHULZE, Dresden; N.N. SIROTA, Moscow
I. TARJAN, Budapest; O.M. TODES, Leningrad
A. A. TSCHERNOV, Moscow
A. F. WITT, Cambridge (U.S.A.)
W. A. WOOSTER, Cambridge

THEORIA CUM PRAXI

Akademie-Verlag · Berlin

Papers should be directed to the editor's office
Scharnhorststraße 20, DDR-7030 Leipzig,
or to the editor Prof. Dr. Hermann Neels,
Zeititzer Weg 25, DDR-7254 Machern,
or to one of the members of the Advisory Board.

Orders can be sent

- in the GDR:
  to the Postzeitungsvertrieb or to the Akademie-Verlag Berlin, Leipziger Straße 3–4, DDR-1086 Berlin;
- in the other socialist countries: to a book-shop for foreign language literature or to the competent news-distributing agency;
- in the FRG and Berlin (West): to a book-shop or to the wholesale distributing agency Kunst und Wissen, Erich Bieber OHG, Wilhelmstr. 4–6, D-7000 Stuttgart 1;
- in the other Western European countries: to Kunst und Wissen, Erich Bieber GmbH, General Wille-Str. 4, CH-8002 Zürich;
- in USA and Canada to the VCH Publishers, Inc., 303 NW 12th Avenue, Deerfield Beach, FL 33442-1788, USA;
- in other countries: to the international book- and journal-selling trade, to Buchexport, Volkseigener Außenhandelsbetrieb der DDR, Postfach 160, DDR-7010 Leipzig, or to the Akademie-Verlag Berlin, Leipziger Straße 3–4, DDR-1086 Berlin.

Journal „Crystal Research and Technology“

Herausgeber: Im Auftrage des Akademie-Verlages Berlin herausgegeben von Prof. Dr. Hermann Neels.

Verlag: Akademie-Verlag Berlin, Leipziger Str. 3–4, DDR-1086 Berlin; Fernruf: 2236221 oder 2236229; Telex-Nr.: 114420; Bank: Staatsbank der DDR, Berlin, Kto.-Nr. 6836-26-20712.

Chefredakteur: Prof. Dr. Hermann Neels, Zeititzer Weg 25, DDR-7254 Machern; Fernruf: Brandis 3442. Anschrift der Redaktion: Scharnhorststr. 20, DDR-7030 Leipzig; Telefon: 310502.

Veröffentlicht unter der Lizenznummer 1526 des Presseamtes beim Vorsitzenden des Ministerrates der Deutschen Demokratischen Republik.

Gesamtherstellung: VEB Druckerei „Thomas Müntzer“, DDR-5820 Bad Langensalza.

Erscheinungsweise: Das Journal „Crystal Research and Technology“ erscheint monatlich. Die 12 Hefte eines Jahrganges bilden einen Band. Bezugspreis je Band 480,– DM zuzüglich Versandspesen; Preis je Heft 40,– DM. Der gültige Jahresbezugspreis für die DDR ist der Postzeitungsliste zu entnehmen.

Bestellnummer dieses Heftes: 1082/22/11.

Urheberrechte: Alle Rechte vorbehalten, insbesondere die der Übersetzung. Kein Teil dieser Zeitschrift darf in irgendeiner Form – durch Photokopie, Mikrofilm oder irgend ein anderes Verfahren – ohne schriftliche Genehmigung des Verlages reproduziert oder in eine von Maschinen, insbesondere von Datenverarbeitungsanlagen verwendbare Sprache übertragen oder übersetzt werden. All rights reserved (including those of translations into foreign languages). No part of this issue may be reproduced in any form, by photoprint, microfilm, or any other means, nor transmitted or translated into a machine language, without written permission from the publishers.
© 1987 by Akademie-Verlag Berlin. Printed in the German Democratic Republic.
AN (EDV) 35830
02800

Cryst. Res. Technol.	22	1987	11	1343—1381

Review

N. N. Sirota

Hydromelioration Institute, Moscow, USSR

Certain Problems of Polymorphism (II)[1]) Generalized Clausius-Clapeyron Equation and Ostwald's Step Rule

Contents

[1]) Part II was read at the crystallography colloquium in Karl-Marx-University, Leipzig, and partly in Humboldt University, Berlin, GDR, in December 1980. Part I: Cryst. Res. Technol. 17, No 6, 661—691 (1982) (hereafter Sirota I).

Introduction

Problems of polymorphic transitions caused by the variation of external generalized forces (pressure, magnetic, electric, gravitational fields, surface energy, etc.), and of origination of new phases from thermodynamically supersaturated initial phase, are of major interest and are playing an ever-increasing role in coping with numerous practical tasks in modern technique and technology. The former president of the Academy of Sciences of the USSR A. P. Alexandrov also paid attention to the actuality of these problems.

The study of the influence of external forces and fields upon relative supersaturation and the kinetics of phase transitions, together with the investigation of mechanisms and kinetics of occurrence of metastable phases in chemical reactions of crystallization, transformations in solids and heat treatment processing of different alloys are directly connected with the problems denoted above and acquire great theoretical and practical interest.

In this context a special principal importance has comprehension, substantiation, interpretation and further development of the famous Ostwald's step rule (OSTWALD) according to which in none of processes the most stable phase is attained at once but first appears the neighbouring condition which is the least stable of all the possible states.

However, now it has become clear that this important rule is not of a general character but is realized under some certain conditions.

Later on we shall consistently consider the relation of generalized Clausius-Clapeyron equation to Ostwald's step rule, and find the factors determining the value of relative thermodynamic supersaturation and the dependence on it of the rate of formation and growth of stable and metastable-phase nuclei. We shall consider conditions of formation of metastable phases in particular in isothermal transformations, and study the influence they exert upon the kinetics of isothermal phase transitions.

Analysis of these questions permits to find quite strictly the limits of Ostwald's step rule applicability.

I. Generalized Clausius-Clapeyron Equation

W. OSTWALD was the first to translate the work of J. W. Gibbs "On the Equilibrium of Heterogeneous Substances" into German, having supplied it with rather interesting comments and a foreword. In fact only after the publication of this translation by Ostwald the works of Gibbs became available to European reader. The publications of J. C. Maxwell also helped in spreading of Gibbs' ideas.

Let's remark that the first translation into Russian appeared in the USSR, edited by V. K. SEMENTCHENKO in 1950.

According to adopted classification based on Gibbs' thermodynamic method, first-order phase transitions are those with non-zero latent heat of transformation, e.g. a transition in one-component substance at constant pressure and constant temperature, when in compliance with the Gibbs' rule there are no degrees of freedom stipulated by external factors. Conversely, non-first (second) order phase transitions under the same external conditions take place without releasing or absorbing latent heat; in their nature they do not correspond to non-variant equilibrium under similar conditions. Formally non-first-order (second order) transitions can be treated as those with additional internal degrees of freedom.

Thermodynamic theory of second-order phase transitions has been studied from different view-points by EHRENFEST, LANDAU (1937), LANDAU and LIFSHITZ, GINZBURG, KHACHATURIAN, VONSOVSKII, SEMENTCHENKO, LEONTOVITCH, K. P. BELOV, SIROTA, SMOLENSKII and others. An original, though incomplete, review has been done by STANLEY. Let's remark that in particular range of thermodynamic supersaturations, the phase transitions of first and second order are similar concerning the kinetics of the processes.

In the study of problems of polymorphism, finding out the conditions of formation and stability of stable and metastable polymorphic modifications the application of Gibbs' thermodynamic method is useful and, possibly, even inevitable.

In case of the first-order phase transition when there are two phases in equilibrium in a one-component system, the variation of Gibbs' free energy is zero[2]), $\Delta Z_{1,2} = 0$, and consequently for the phases 1 and 2

$$dZ_1 = dZ_2 .$$

Gibbs free energy is a characteristic function of generalized forces X_i — temperature, pressure, magnetic, and electric fields — and the mole number n_j,

$$Z = Z(T, X_i, n_j) \equiv Z(T, p, \mathcal{H}, \mathcal{E}, n, \ldots) ,$$

and so

$$dZ_1 = \left(\frac{\partial Z_1}{\partial T}\right) dT + \sum_i \frac{\partial Z_1}{\partial X_i} dX_i + \sum_j \frac{\partial Z_1}{\partial n_j} dn_j ,$$

$$dZ_2 = \left(\frac{\partial Z_2}{\partial T}\right) dT + \sum_i \frac{\partial Z_2}{\partial X_i} dx_i + \sum_j \frac{\partial Z_2}{\partial n_j} dn_j ,$$

where $\dfrac{\partial Z}{\partial X_i} = x_i$ are generalized coordinates[3]), $\dfrac{\partial Z}{\partial n_j} = \mu_j$ are chemical potentials. From these equations for one-component substance with constant mole number, $n_j = \text{const}$ we infer that the derivative of temperature with respect to any other generalized force X_q along equilibrium line is:

$$\frac{dT}{dX_q} = - \frac{\dfrac{\partial(Z_1 - Z_2)}{\partial X_q} + \sum\limits_{i \neq q} \dfrac{\partial(Z_1 - Z_2)}{\partial X_i} \dfrac{\partial X}{\partial X_i}}{\partial(Z_1 - Z_2)/\partial T} .$$

This equation may be considered as generalized Clausius-Clapeyron equation.

In contrast to the above equation when many of the generalized forces are changing simultaneously (e.g. temperature varies together with pressure and magnetic field intensity) the study of the functional dependences is overcomplicated.

However, quite clear consideration is possible to be conducted in different sections of the multi-dimensional space when all the generalized forces but two of them are constant.

For example, when T and X_q are changing and all the other forces are constant ($X_{i \neq q} = \text{const}$, $dX_{i \neq q} = 0$), then

$$\frac{dT_k}{dX_q} = - \frac{x_{1q} - x_{2q}}{S_1 - S_2} .$$

---

[2]) Presently Gibbs' free energy is as a rule denoted G; we conserve the notation Z as in part I, which was earlier adopted in German.

[3]) Such are, for example, entropy S, volume V, electric $\left(\dfrac{1}{4\pi} D\right)$ and magnetic induction $\dfrac{1}{4\pi} B$, etc. (e.g. see SEMENTCHENKO, 1968; BAZAROV).

Here $S_1 - S_2$ is the difference of entropies of phases 1 and 2 at $T = T_k$, equal to the difference of enthalpies divided by the transition temperature:

$$S_1 - S_2 = \frac{H_1 - H_2}{T_k},$$

then

$$\frac{dT_k}{dX_q} = -T_k \frac{x_{1q} - x_{2q}}{H_1 - H_2},$$

or

$$\frac{d \ln T_k}{dX_q} = -(x_{1q} - x_{2q})/(H_1 - H_2)\,.$$

When pressure is changing ($X_q = p$, $x_q = V$), the above equation acquires the form:

$$\frac{d \ln T_k}{dp} = -(V_1 - V_2)/(H_1 - H_2)_p\,.$$

When magnetic field strength is changing $\left(X_q = \mathcal{H},\ x_q = -\frac{\mathcal{H} + 4\pi\mathcal{M}}{4\pi}\right)$,

$$\frac{d \ln T_k}{d\mathcal{H}} = (M_1 - M_2)/(H_1 - H_2)\,.$$

When the electric field strength is changing $\left(X_q = \mathcal{E},\ x_q = -\frac{\mathcal{E} + 4\pi P}{4\pi}\right)$,

$$\frac{d \ln T_k}{d\mathcal{E}} = (P_1 - P_2)/(H_1 - H_2)\,.$$

Here P is polarization, M is magnetization. When two of the generalized forces X_p and X_q are varying but temperature T, together with all the other forces, is constant then

$$\frac{dX_p}{dX_q} = -\frac{\dfrac{\partial(Z_1 - Z_2)}{\partial X_q}}{\dfrac{\partial(Z_1 - Z_2)}{\partial X_p}} = (x_{1q} - x_{2q})/(x_{1p} - x_{2p})\,.$$

For example,

$$\frac{dp_k}{d\mathcal{H}} = \frac{M_1 - M_2}{V_1 - V_2}; \quad \frac{dp_k}{d\mathcal{E}} = \frac{P_1 - P_2}{V_1 - V_2}; \quad \frac{\partial \mathcal{H}_k}{\partial \mathcal{E}} = -\frac{P_1 - P_2}{M_1 - M_2}$$

Let us analyse the behaviour of the Gibbs free energy curves of the plases 1, 2 near their mutual equilibrium using the methods of geometrical thermodynamics supposing that external generalized forces are changing (MLODZEYEVSKII 1922, 1937, 1948, 1956; SIROTA 1968). Gibbs free energy of any phase is a characteristic function of generalized forces, including temperature, and a mole number n_j of each component, so that

$$dZ = -S\,dT + \sum x_i\,dX_i + \sum_j \mu_j\,dn_j\,.$$

In case of one-component substance

$$dZ = -S\,dT + \sum_i x_i\,dX_i\,.$$

When one of the generalized forces X_i is changing this equation describes the peculiarities of the free energy curve as a function of this force.

The total entropy S is always positive and so its first-order derivative is $dS/dT = c_p/T > 0$. Consequently, $dZ/dT < 0$. The second-order derivative $\partial^2 Z/\partial T^2 < 0$, so the curve Z versus T is falling, its convexity being directed off the abscissa axis.

Quite similar analysis with respect to other generalized forces shows that

$$\frac{\partial Z}{\partial p} > 0; \quad \frac{\partial^2 Z}{\partial p^2} < 0; \quad \frac{\partial Z}{\partial \mathcal{H}} < 0; \quad \frac{\partial^2 Z}{\partial \mathcal{H}^2} < 0; \quad \frac{\partial Z}{\partial \mathcal{E}} < 0; \quad \frac{\partial^2 Z}{\partial \mathcal{E}} < 0 .$$

II. Change of the generalized coordinates in the phase transition in equilibrium. Thermodynamic supersaturations caused by the deviation of the generalized forces from their equilibrium values

When one of the generalized forces deviates from thermodynamic equilibrium, corresponding, e.g., to non-variant two-phase equilibrium of one-component system, a certain thermodynamic supersaturation is achieved, characterized by the magnitude of the difference $\Delta Z_{12} = Z_1 - Z_2$ of specific Gibbs free energies of phases 1, 2.

In case of non-variant two-phase thermodynamical equilibrium of one-component system $\Delta Z_{12} = 0$. In particular, when $p =$ const the change of specific entropy $\Delta S_{12} = S_1 - S_2$ in this case is equal to the ratio of phase transition heat $\Delta H_{12} = H_1 - H_2$ to the equilibrium temperature T_k:

$$\Delta S = \frac{\Delta H_{12k}}{T_k} = \frac{\Delta H_{012}}{T_k} + \frac{1}{T_k} \int_0^{T_k} (C_{p1} - C_{p2})\, dT = \int_0^{T_k} \frac{C_{p1} - C_{p2}}{T}\, dT .$$

Note that entropy and enthalpy of high-temperature phase are always higher than corresponding quantities for low-temperature phase,

$$S_2 > S_1; \qquad H_{2k} > H_{1k} .$$

For the two-phase equilibrium of one-component substance at $T =$ const, $p = p_k$, $\Delta Z_{12} = 0$ and the difference of specific volumes $\Delta V_k = V_1 - V_2$ is non-zero. In the situation of transition induced by the change of the pressure, the phase 1 of lower pressure, stable at $p \leqq p_k$ has in accordance with Le Chatelier rule, specific volume greater than the higher-pressure phase 2, stable at $p \geqq p_k$, i.e.

$$V_1 = (\partial Z_1/\partial p)_{pk} > V_2 = (\partial Z_2/\partial p)_{p_k} .$$

In this case, as $p_k(V_1 - V_2)_k = -\Delta H_k$ when $T = T_k$, $p = p_k$,

$$\Delta V_{12k} = -\frac{\Delta H_k}{p_k}.$$

In the general case of one-component substance, phase transition being induced by the change of any of the generalized forces X_i in equilibrium $\Delta Z_{12} = 0$ and, taking into account that $\frac{\partial Z}{\partial X_i} = x_i$, the equation holds:

$$(x_{1i} - x_{2i}) = -\frac{\Delta H_{12k}}{X_{ik}}$$

or

$$(\bar{\chi}_{1i} - \bar{\chi}_{2i})_k = \Delta H_k / X_{ik}^2 ,$$

where $(x_{1i} - x_{2i})_k / X_{ik} = (\bar{\chi}_{1i} - \bar{\chi}_{2i})_k$ is the difference of generalized mean susceptibilities of phases 1, 2. Here $\Delta H_k \equiv \Delta H_{12k}$ is the enthalpy of the phase transition caused by the change of the generalized force X_i taken at its equilibrium value X_{ik}.

From this follows that in equilibrium of one-component substance (characterized by T_k, p_k or X_k) the change of entropy is equal to the product of the difference of generalized coordinates by the ratio of varying generalized force to the transition temperature,

$$\Delta S_k = -\frac{\Delta x_{ik}}{T_k} X_{ik} .$$

In particular, when three of the quantities characterizing equilibrium are known, this relation permits to find the fourth quantity.

The value of thermodynamic supersaturation of one-component substance, when one of the generalized forces X_i deviates from equilibrium value X_{ik}, equals to

$$\Delta Z = \Delta Z_k + \int \left(\frac{\partial\, \Delta Z_{12}}{\partial X_i}\right) dX_i + \dots .$$

When the deviation from equilibrium is relatively small, the magnitude of thermodynamic supersaturation ΔZ may be represented as a series (SIROTA 1971):

$$\Delta Z_{12} = \left(\frac{\partial \Delta Z_{12}}{\partial X_i}\right) \Delta X_{ik} + \left(\frac{\partial^2\, \Delta Z_{12}}{\partial X_i^2}\right) \frac{\Delta X_{ik}^2}{2} + \dots$$

$$\Delta Z_{12} = (x_{1i} - x_{2i})\,(X_i - X_{ik}) + \left[\frac{\partial (x_{1i} - x_{2i})}{\partial X_i}\right] \frac{(X_i - X_{ik})^2}{2 + \dots} + \dots$$

Undoubtedly the subject of substantial interest is the determination of thermodynamic supersaturation value not only with respect to initial equilibrium phase, but to all possible non-equilibrium phases (modifications) of a given substance. To evaluate the magnitude of thermodynamic supersaturation we may use formulae given above, substituting the equilibrium magnitudes of generalized forces X_k by their values for quasi-equilibrium transitions, corresponding to metastable phases.

Leaving only the first term of the series, we find that specific thermodynamic supersaturation of one-component substance equals to the product of the deviation of the generalized force X by the difference of conjugate specific generalized coordinates of phases 1, 2. Using the value of this product given above we find:

$$\Delta Z = (x_{1i} - x_{2i})\,(X_{ik} - X_i) \cong -\Delta H_k (\overline{\chi}_{1i} - \overline{\chi}_{2i})\, X_i X_{ik} \cong -\frac{X_{ik} - X_i}{X_{ik}} \Delta H_k .$$

Note that

$$X_i X_{ik} \approx X_{ik}^2; \qquad \Delta H_k \equiv \Delta H_{12k} .$$

Thus thermodynamic supersaturation is equal to: in case of undercooling:

$$\Delta Z_T = -(S_1 - S_2)\,(T_k - T) \cong -\Delta H_k + \left(\frac{C_{p1}}{T_k} - \frac{C_{p2}}{T_k}\right) T_k^2 \cong \frac{\Delta H_k (T_k - T)}{T_k};$$

when pressure is changing:

$$\Delta Z_p = (V_1 - V_2)\,(p_k - p) \cong -\Delta H_k - (\overline{\chi}_{p1} - \overline{\chi}_{p2})\, p_k^2 \cong -\frac{p_k - p}{p_k} \Delta H_k;$$

when magnetic field strength is changing

$$\Delta Z_{\mathcal{H}} = (M_1 - M_2)\,(\mathcal{H} - \mathcal{H}_k) \cong -\Delta H_k + (\overline{\chi}_{\mathcal{H}1} - \overline{\chi}_{\mathcal{H}2})\, \mathcal{H}^2 \cong \frac{\mathcal{H}_k - \mathcal{H}}{\mathcal{H}_k} \Delta H_k;$$

when electric field strength is changing

$$\Delta Z_{\mathcal{E}} = (P_1 - P_2)\,(\mathcal{E} - \mathcal{E}_k) \cong -\Delta H_k + (\overline{\chi}_{\mathcal{E}1} - \overline{\chi}_{\mathcal{E}2})\, \mathcal{E}_k^2 \cong \frac{\mathcal{E}_k - \mathcal{E}}{\mathcal{E}_k} \Delta H_k ,$$

where M is intensity of magnetization, P — polarization, $\chi_i = \frac{\partial x_i}{\partial X_i}$; $\chi_i = \frac{x_i}{x_i}$ true and average generalized susceptibilities.

As follows from the right sides of the given equations, the values of thermodynamic supersaturation for one-component substances to a first, linear approximation are equal to the difference of enthalpies plus the product of the difference of average generalized susceptibilities by the square of the conjugate generalized force. On the other hand

$$\Delta Z = -\Delta H_k \frac{\Delta X_k}{X_k} = -\Delta S \frac{\Delta X_k}{X_k} T_k ,$$

i.e. the value of specific thermodynamic supersaturation is proportional to the product of the transition heat by relative deviation of the generalized force from the non-variant equilibrium value.

III. Thermodynamic supersaturation in two- and multi-component systems

Thermodynamic supersaturation in two and multi-component systems may be calculated in geometric thermodynamic approximation of the regular-solution theory, in quasi-chemical and other approximations.

Using the methods developed earlier (Becker; Borelius; Pinés 1941, 1961; Danilov, Kamenetskaya; Sirota (1944, 1948, 1955, 1970); Ivanov), based on a semi-empirical interpretation of Roseboom and Van-Laar's method (see also Urazov; Tammann 1924; Vogel; Mlodzeyevskii 1922, 1947; Anosov, Pogodin; Anosov et al.; Dehlinger; Hume-Rothery and Reynor; Darken and Gurry; Lumsden; Paufler; Paufler, Lenchner) one can find quantitatively the elements of phase diagram — liquidus and solidus lines and surfaces, limits of existence of solid solutions — and also determine the magnitude of thermodynamic supersaturation in case of different undercoolings, composition changes, or when generalized forces deviate from their equilibrium values.

Temperature dependence of the concentration limit of existence of solid solution of α in the basic component A, from which, as the temperature falls, chemical compound $A_mB_n(\gamma)$ with concentration C_γ is precipitated, may be determined from phase equilibrium condition, i.e. equality of chemical potentials of phases α and γ at given temperature and pressure:

$$\mu_A^\alpha = \mu_A^\gamma; \qquad \mu_B^\alpha = \mu_B^\gamma .$$

This equilibrium condition is equivalent to a system of equations:

$$\frac{\partial Z_\alpha}{\partial C} = \frac{\partial Z_\gamma}{\partial C}; \quad Z_\alpha - Z_\gamma = (C_\gamma - C_\alpha) \frac{\partial Z_\alpha}{\partial C} .$$

In the approximation of the theory of solutions with different degrees of regularity (Guggenheim; Sirota; Christian) the free energy of two-component solid solution α is

$$Z_\alpha = Z_A + (Z_B - Z_A) C_\alpha + K_\alpha C_\alpha (1 - C_\alpha) + RT[C_\alpha \ln C_\alpha + (1 - C_\alpha) \ln (1 - C_\alpha)] .$$

In general in this case interaction coefficient of the components in the solution, K_α depends on temperature and composition (Pinés; Sirota 1948, 1970). Thus

$$K_\alpha = K'_{0\alpha}(1 + \varkappa_\alpha T)(1 + \lambda C_\alpha) \cong K'_{0\alpha}[1 + \lambda C_\alpha] a = K'_{0\alpha}(P + QT) = \\ = K_{0\alpha}(1 + \lambda C_\alpha) .$$

Here

$$a = 1 + \varkappa_\alpha T; \qquad P = 1 + \lambda_\alpha C_\alpha; \qquad Q = \varkappa_\alpha P .$$

Neglecting concentrational and temperature dependences of K_α we have:

$$\frac{\partial Z_\alpha}{\partial C} = (Z_B - Z_A) + K_\alpha(1 - 2C_\alpha) + RT \ln \frac{C_\alpha}{1 - C_\alpha}.$$

Proceeding from the phase equilibrium equation and considering C_γ to be a constant, when solubility is small, $C_\alpha \to 0$, we obtain an approximate expression (assuming that components don't dissolve in compound γ):

$$\ln C_\alpha = -[K_\alpha + Z_B - Z_A + 1/C_\gamma(Z_A - Z_\gamma)]/RT .$$

The corresponding supersaturations are

$$\Delta Z_{\Delta T} \cong RT_1 \ln C_\alpha(T_2)/C_\alpha(T_1) .$$

At present this method of geometric thermodynamics is widely used for different computer calculations of phase diagrams (see. e. g. KAUFMAN, BERNSTEIN).

Let us consider another example: the existence of continuous solid solution and its spinodal decay.

When interaction coefficient is negative, $K_\alpha < 0$, continuous solid solution will be thermodynamically stable at all temperatures, whereas in the entire concentration region will be $K_\alpha < 0$ and the curve free energy versus composition will not have inflection points and several extrema.

When the coefficient of interaction of the components in solid phase is positive, $K_\alpha > 0$, lower than a certain temperature T_k two-phase decay of the solid solution into phases α_1, α_2 takes place; the concentration boundaries of the decay in symmetrical case are described by the equation

$$RT \ln \frac{1 - C_{\alpha 2}}{1 - C_{\alpha 1}} = K_{0\alpha}[(\lambda - 1)(C_{\alpha 2}^2 - C_{\alpha 1}^2) - 2\lambda(C_{\alpha 2}^3 - C_{\alpha 1}^3)] .$$

Hence, using L'Hopital rule, at $C_{\alpha 1} = C_{\alpha 2} = C_{ak}$ we find the critical temperature: $T_k = \frac{2K_0}{R}[(\lambda - 1) - 6\lambda C_{\alpha k}](1 - C_{\alpha k})^2$; at higher temperatures continuous solid solution exists.

The boundary of heterogeneous region at $C_{\alpha 1} \to 0$ in this case is described by the equation:

$$\ln(1 - C_{\alpha 2}) = -K_{0\alpha}(\lambda + 1)/(RT) .$$

This equation is in its form similar to that of SHREDER, derived and analysed by ANOSOV (1953).

Under this condition the value of thermodynamic supersaturation of the solid solution at the temperature falling from T_2 to T_1 is defined by the expression (concentrations are supposed to be small and phase boundary is plane):

$$\Delta Z_{21} = K_{0\alpha}(\lambda + 1)(T_2 - T_1)/T_2 .$$

IV. Thermodynamic supersaturation caused by light, radiation and other external actions

A problem of special interest is the study of residual influence of penetrating radiation (electromagnetic, light, ultrasonic irradiation) and also the influence of residual cold plastic deformation upon the value of thermodynamic supersaturation.

Penetrating radiation action, plastic deformation and other kinds of excitation create, as a rule, non-uniformly distributed state with an increased value of Gibbs free energy. All sorts of excitation cause non-uniformity of state also because of different local defects each of them having its relaxation time.

If we confine ourselves only to mean values of Gibbs free energy, then averaged value of thermodynamical supersaturation at any temperature is determined by specific absorbed energy in the process of external action. In one-component substances it leads to the change of phase transition temperature (see SIRCTA I).

In case of limited-region solid solution subject to external action, the increase of its specific Gibbs free energy causes the shift of miscibility gap (i.e. change of solubility of isolating surplus phase, SIROTA 1955, 1961; BALARIN, SCHENK). When mainly initial solution is excited, the miscibility gap shifts to smaller concentrations. On the contrary, when irradiation or other external action is applied to heterogeneous mixture of initial solid solution α and isolating surplus phase (solution or composition) γ, the latter's specific energy increases to a greater extent than that of the phase α; in this case miscibility gap shifts to greater concentration, i.e. solubility of the phase γ in the phase α increases.

When solid solution is subjected to spinodal decay the increase of its free energy by external action leads to the widening of concentration region where the solutions coexist; in this case metastable compound γ' may isolate.

Figures 1a and 1b schematically illustrate situations when as a result of irradiation and other external actions the free energy of phase 1 and correspondingly, phase 2 mainly increases; also shown is the growth of supersaturation versus composition in the cases considered.

As a result of non-uniformity of the excited state, a major local increase of specific free energy may stipulate a "trigger" mechanism facilitating the occurance of the centre of a new phase in microscopic volumes with increased thermodynamic supersaturation, since local supersaturation may exceed a threshold necessary for the nuclei of critical size to appear and thus stimulate the process of phase transition. On the other hand, local excitation of the forming nuclei of a new phase may lead to their dissolution.

V. Thermodynamic supersaturation with respect to stable and metastable phases

Simple substances and compounds which may be regarded as one-component ones, may exist not only in stable, lowest energy structural states, but also in several additional, higher excited metastable states.

Near absolute zero of temperature at normal pressure the minimum energy (ground) state corresponds to the highest energy of atomization H_{01}. All other metastable states have lower value of H_0:

$$|H_{01}| > |H_{0a1}| > \dots > |H_{0an}| .$$

As a result of polymorphic transitions metastable phases (with respect to low-temperature phase) may become stable either at increased pressures or at sufficiently high temperatures, e.g. near the melting point, or even, in principle, at higher temperatures (overheating).

Taking into account this peculiarity of such metastable modifications, supposing that the difference of their free energies and that of a stable phase is small, we con-

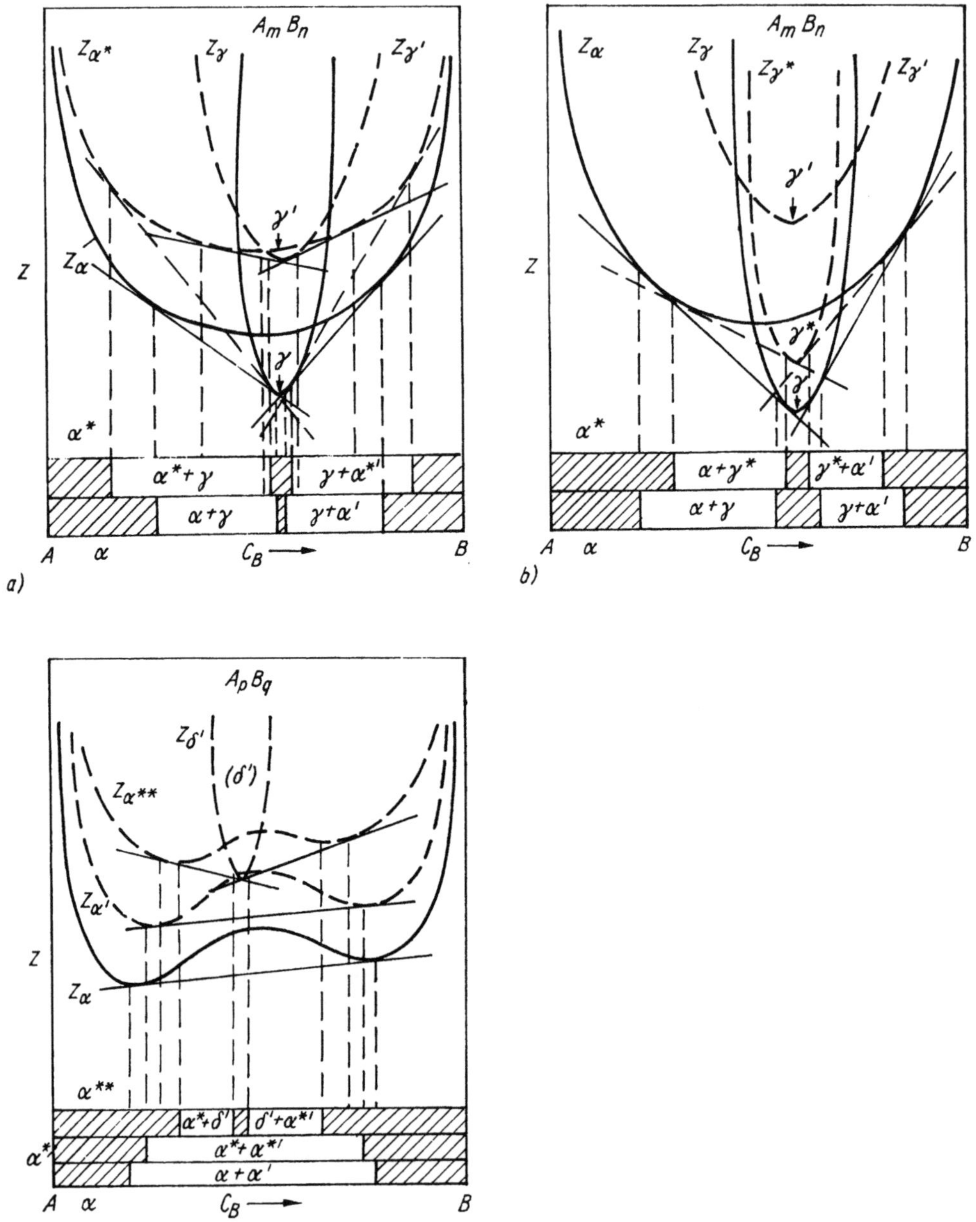

Fig. 1. Change of equilibrium conditions in binary system in case of excitation: a — of solid solution α, b — of stable compound $A_mB_n(\gamma)$, c — of solid solution subject to spinodal decay. Curves of Gibbs' free energy Z_α, Z_α^* refer to solid solution α, Z_γ, Z_γ^* — to compound A_mB_n in normal and excited states; Z_γ refers to metastable modification $A_mB_n(\alpha')$, Z_γ' — to metastable compound $A_pB_q(\delta')$

clude that to a first approximation the curves of $Z_{a1}, \dots Z_{an}$ versus temperature are practically parallel to the curve of the free energy Z_1 of equilibrium phase[4]).

In this case the value of thermodynamic supersaturation with respect to phase 2 at temperatures lower than T_k (which is the temperature of equilibrium transition of phase 1 into phase 2) will be determined by the magnitude of undercooling with respect to the temperatures $T_{a1}, \dots, T_{an}$, corresponding to the intersection points of the curves of the free energies of phases $a_1, \dots, a_n$ with that of the initial phase 2. Then

$$T_{an} = T_{k12} - \frac{H_{021} - H_{02an}}{S_2 - S_1}; \qquad T_{an} = T_{k12} \frac{\Delta H_{02an}}{\Delta H_{021}}.$$

The value of thermodynamic supersaturation of phase 2 with respect to the stable phase 1 and metastable phase a_n at temperature T may be evaluated by relation (see Fig. 2):

$$\Delta Z_{2an} = \Delta H_{021}(T_{2an} - T)/T_k ,$$

where ΔZ_{2an} is supersaturation with respect to metastable phase at; ΔZ_{21} is supersaturation with respect to stable phase 1, T is the temperature, lying lower than the

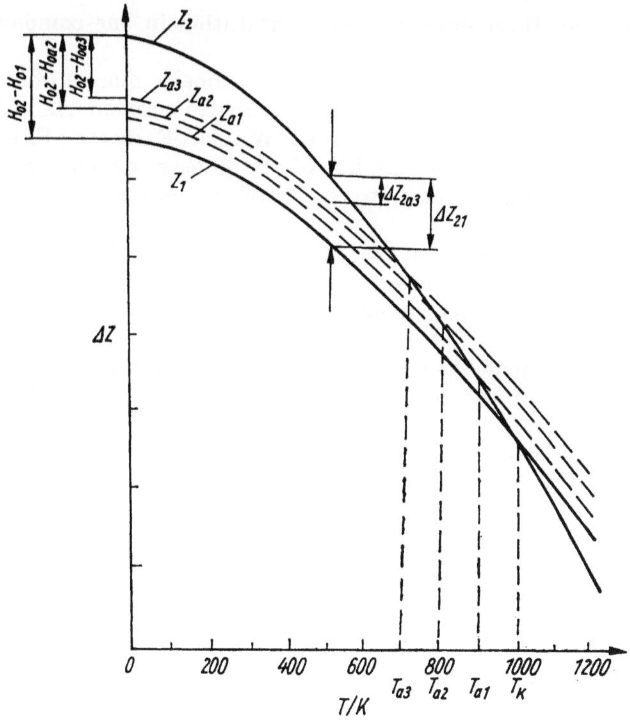

Fig. 2. Curves Gibbs free energy versus temperature for the phases: 2 (Z_2) and 1 (Z_1); metastable phases a_1 (Z_{a1}), a_2 (Z_{a2}), a_3 (Z_{a3})

---

[4]) In reality the decrease of atomization energy $|H_0|$ leads to the increment of entropy, i.e. rate of decrease of $-(\partial Z/\partial T)$ with the growth of temperature. The condition of parallelism of the curves may be adopted only as a first approximation when the deviations $H_{02}-H_{0an}$ of metastable phases atomization energy from that of a stable phase are small.

temperature of quasi-equilibrium transition from initial phase 2 to metastable phases *an*, so that

$$T_{an} < T_k; \qquad T < T_{an} .$$

When some generalized force X deviates from its equilibrium magnitude X_k, the values of thermodynamic supersaturation when the above suppositions are valid (i.e. Gibbs free energy of metastable phase is near that of phase 1 and both curves are parallel) are approximately as follows:

$$\Delta Z_{2an} = \Delta H_{021}\left(\frac{\Delta H_{02n}}{\Delta H_{021}} - \frac{T}{T_k}\right)$$

or

$$\Delta Z_{2an} = \Delta H_{21}(T_{an} - T)/T_k ,$$

adopting $\Delta H_{021} = \Delta H_{21}$.

Already this linear approximation in many cases appears to be sufficient for finding out of conditions and explanation of peculiarities of appearance of metastable phases in one-component systems depending on thermodynamic supersaturation.

VI. Work on appearance of nuclei of stable and metastable phases depending on thermodynamic supersaturation in one-component systems

For appearance and growth of nuclei of thermodynamically more stable phases than initial one, it is necessary that thermodynamic supersaturation ΔZ should exceed a certain critical value (GIBBS; VOLMER; VOLMER, WEBER; FARKASH; FRENKEL, 1939). The latter depends on the degree of stability of the phases, on energetic properties of initial and originating phases and, to a certain extent, on the transformation mechanism.

We proceed from the idea that whatever is the mechanism, the transformation has probabilistic character, each mechanism having its specific activation energy for transfer of atoms into positions corresponding to the centres of formation and growth of a new phase.

There are so-called homogeneous and heterogeneous crystallization processes. The first take place in absence of alien inclusions or substrates, the second in presence of them.

The idea that heterogeneous crystallization processes are those with ready centres of nucleation has a relative meaning. The formation of new phase nuclei on the bases or on the surfaces of inclusions is also a probabilistic process as in case of their spontaneous formation. There is a certain probability that a two-dimensional nucleus of new phase should appear and grow on the surface of a substrate or seed.

It is necessary to stress that in heterogeneous crystallization processes the value of thermodynamic supersaturation is considerably smaller than in homogeneous ones.

The effective activity of the substrate depends on microstructure of its surfaces, its crystallochemical affinity for the new phase, and on the other hand, on its macrostructure, i.e. presence of convex steps in the relief, cracks, recesses and pores (LEMLEIN; DANKOV; BLIZNAKOV; GORBUNOVA 1935, 1957; SIROTA 1962; FRENKEL; TEMKIN).

Let's remark that sometimes in recesses and pores the effective thermodynamic supersaturation required for the origination of new-phase nuclei is quite small or even zero (KAZACHKOVSKII; SIROTA 1962; TURNBULL 1950). Note also that new-phase growth on the surface of the base may take place via step dislocational mechanism (LEMLEIN; LEMLEIN, DUKOVA; FRANK, VAN DER MERVE; FRANK; FRANK, READ;

BURTON et al.; CHERNOV, A. A.) or via mechanism of "normal growth" (BARTENEV; REISS; BORISOV, DUKHIN; IVANTSOV, TILLER).

Adequate understanding of the mechanism of heterogeneous crystallization process in presence of bases and seeds, and condition of formation of metastable polymorphic modifications and phases is possible only on the basis of the principle of dimensional and orientational similarity of the crystal structure of initial and appearing phases and the base. This principle was elaborated by Dankov (DANKOV's principle), and then by BLIZNAKOV, ARKHAROV, TURNBULL and a.o.

In a wider sense it is necessary to take into account not only the crystallographic similarity of the phases taking part in the crystallization process, but also macrostructure of the base, wetting angles which has been shown already by VOLMER (1929), and the presence of near order in amorphous bases. (On the crystal-chemical nature of the phases see e.g. N. V. BELOV, BOKII, KREBS, KLEBER, ZHDANOV, PAUFLER and LENCHNER a.o.)

Numerous examples of crystallographic similarity of initial and appearing phases and a base were given by VOLMER and FLOOD, by BLIZNAKOV, KAISCHEW, MEHL, KURDJUMOV and SACHS, and by WASSERMANN.

As follows from above, relative thermodynamic supersaturation $\Delta Z/Z_k$ (relative undercooling $\Delta T/T_k$, relative pressure variations $\Delta p/p_k$ or $\Delta X_i/X_i$ when $T < T_k$, $p < p_k$, $X < X_k$) is the greatest with respect to the most stable phase and smallest with respect to the least stable of all phases, formation of which is thermodynamically possible under given conditions:

$$\frac{T_k - T}{T_k} > \frac{T_{a1} - T}{T_{a1}} > \dots > \frac{T_{an} - T}{T_{an}}; \quad \frac{p_k - l}{p_k} > \dots > \frac{p_{an} - p}{p_{an}};$$

$$\frac{X_{ik} - X_i}{X_{ik}} > \dots > \frac{X_{ian} - X_i}{X_{ian}}.$$

Consideration of the shapes of free energy curves for initial and metastable phases illustrates that when supersaturation is small at $T_a < T < T_k$, and $p_d < p < p_k$ or in general $X_a < X < X_k$ only stable phase 1 is thermodynamically possible and formation of other phases is in principle prohibited,

Maximum work on appearance of nuclei (three-dimensional nuclei — centres of crystallization and two-dimensional nuclei addition of which conditions the rate of new phase growth) depends not only on the value of thermodynamic supersaturation ΔZ, but also on surface energy σ at the phase boundary, elastic deformation energy E_g caused by volume change[5]) and also work of other generalized forces related to the generalized coordinates $E_{xi} = \int X_i d(\Delta x_i)$. The change of Gibbs free energy $\Delta\Psi_3$ with the appearance of a three-dimensional nuclei with volume V_3 of new phase in thermodynamically supersaturated initial phase of one-component system is

$$\Delta\Psi_3 = -\frac{\Delta Z}{V} v_3 + \frac{v_3}{V}[E_g + E_{\varkappa} + E_{\mathfrak{E}} + \dots + T\,\Delta S_e] + K_{S3}\sigma' v_3^{2/3} =$$

$$-\frac{\Delta Z'}{V} v_3 + K_{S3}\sigma' v_3^{2/3}.$$

[5]) One of the first rather exact calculations of deformation energy in the process of formation of a new phase has been complied by MOTT and NABARRO, NABARRO (in anisotropic media see CAHN). The additional work resulting from volume change, electric and magnetic field change in the expressions for the work on appearance of nuclei and the rate of their formation — has been primarily accounted for by SIROTA (1945, 1946, 1948c, d, 1957, 1961a, b, 1962).

Here ΔZ is mole free energy difference for initial and appearing phases, $-\Delta Z' = -\Delta Z + \sum_j E_j - T\,\Delta S_e$; ΔS_e is entropy change per mole as a result of new phase formation, $\frac{1}{V}\sum E_j = \frac{1}{V}[E_g + E_{\varkappa} + E_g^{+}]$ is total energy change per unit volume as a result of the change of the volume, electric, and magnetic field intensity, correspondingly. K_{S3} is a form factor. In case of a sphere $K_{S3} = (36\pi)^{1'3}$, in case of a cube $K_{S3} = 6$.

The equation $\frac{\partial \Psi_3}{\partial V_3} = 0$ gives the critical volume of a three-dimensional nucleus:

$$v_3 = (2/3K_{s3})^3\,\sigma'^3/\Delta Z'^3 V^3 .$$

Reduced size of a three-dimensional nuclei of the cubic shape with a side a_3 is $a_3 = \frac{4V\sigma'}{\Delta Z'}$ for a sphere $r_3 = \frac{2V\sigma'}{\Delta Z}$; if $\sum E_j = 0$ then $r_3 \cong \frac{2V\sigma'}{\Delta H}\frac{T_k}{\Delta T}$; because $\Delta Z = \Delta H\,\Delta T/T_k$. Number of atoms (molecules) in the nucleus of critical size is then

$$n_3 = \frac{v_3}{v_A} = \left(\frac{2}{3}K_{S3}\right)^3\left(\frac{\sigma'}{\Delta Z'}\right)^3 N^3 v_A^2$$

where v_A is the atomic (molecular) volume. For a spherical nucleus

$$n_3 = \frac{32}{3}\pi\frac{N^3\sigma'^3 v_A^2}{\Delta Z'^3}\,;\quad \text{if}\quad \sum E_j = 0\,,$$

then

$$n_3 = \frac{32}{3}\pi\frac{N^3\sigma'^3 v_A^2}{\Delta H^3}\left(\frac{T_k}{\Delta T}\right)^3.$$

Work on the formation of a critical-size nucleus is

$$A_3 = \frac{1}{2}\left(\frac{2}{3}K_{S3}\right)^3\frac{\sigma'^3}{\Delta Z'^2}V^2;$$

if

$$\sum_j E_j = 0 \quad \text{then} \quad A_3 = \frac{1}{2}\left(\frac{2}{3}K_{S3}\right)^3 V^2\frac{\sigma'^3}{\Delta H^2}\left(\frac{T_k}{\Delta T}\right)^2.$$

In case of as cubic nucleus

$$A_3 = 32V^2\sigma^{3'}/\Delta Z'^2;$$

if

$$\sum E_j = 0 \quad \text{then} \quad A_3 = 32V^2\frac{\sigma'^3}{\Delta H^2}\left(\frac{T_k}{\Delta T}\right)^2.$$

Effective value of surface energy σ' depends on the size of a chip, a drop; e.g., according to TOLMAN, and to KIRKWOOD and BUFF $\sigma' = \frac{\sigma_0}{1+2\sigma_0/r}$ or the smaller the nuclei the stronger the dependence of the effective value of σ' on the size r. On the other hand, σ' depends on elastic tensions and deformation of surface layers, and on gradients of generalized forces.

Analogically to the idea of VOLMER, KOSSEL-STRANSKI-KAISCHEW of the crystal growth mechanism by the addition of two-dimensional nucleus, the change of Gibbs free energy in this process is

$$\Delta\Psi_2 = -\frac{\Delta Z}{V} v_2 + \frac{v_2}{V}[E_g + E_\varkappa + E_\varepsilon + \ldots - T\,\Delta S_e] = -\frac{\Delta Z'}{V} v_2 + K_{S2}\varkappa v_2^{1/2}$$

where K_{S2} is a formfactor of two-dimensional nucleus. In case of a disc (of thickness δ) $K_{S2} = 2\sqrt{\pi/\delta}$; in case of square prism $K_{S2} = 4/\sqrt{\delta}$; $\varkappa$ is peripheral energy.

From the equation $\frac{\partial\Psi_2}{\partial v_2} = 0$ critical volume of a two-dimensional nucleus may be found:

$$v_2 = \left(\frac{K_{S2}\varkappa V}{2\Delta Z'}\right)^2; \quad \text{if} \quad \sum_j E_j = 0 \quad \text{then} \quad v_2 = \left[\frac{K_{S2}V\varkappa}{\Delta H}\frac{T_k}{\Delta T}\right]^2.$$

The side of a square nucleus of critical size a_2 is

$$a_2 = \frac{2V\varkappa}{\Delta Z'\delta} = \frac{2V\sigma}{\Delta Z'}; \quad \text{if} \quad \sum E_j = 0 \quad \text{then} \quad a_2 = \frac{2V\sigma}{\Delta H}\frac{T_k}{\Delta T}.$$

Radius of the disc nucleus is $r_2 = \frac{V\varkappa}{\Delta Z'\delta} = V\sigma/\Delta Z'$;

if

$$\sum_j E_j = 0 \quad \text{then} \quad r_2 = \frac{V\sigma}{\Delta H}\frac{T_k}{\Delta T}.$$

Here peripheral energy $\varkappa$ is proportional to the surface energy, $\varkappa \cong \delta\sigma \cong d\sigma$ (here d is interplane distance), according to STRANSKI and KAISCHEV. The work on formation of two-dimensional nucleus of a critical size is $A_2 = \left(\frac{K_{S2}}{2}\right)^2 \frac{V\varkappa^2}{\Delta Z'}$; so for a square nucleus $A_2 = \frac{4V\varkappa^2}{\Delta Z'\delta}$; for a disc nucleus $A_2 = \frac{\pi V\varkappa^2}{\Delta Z'\delta}$.

A. Work on formation of three-dimensional nuclei depending on thermodynamical supersaturation caused by the temperature change

Let us consider the temperature dependence of the work on nuclei formation. We adopt a number of simplifying assumptions: i) absence of external fields and stresses at the phase boundaries of one-component substance, $\sum_j E_j = 0$; ii) proportionality of surface energy at the phase boundaries to specific energy of phase transition, $\sigma = \beta\,\Delta H$. (The question of proportionality of surface energy at $T = 0$ to atomization heat and to melting-heat follows from the analyses of STRANSKI and KAISCHEW, of SIROTA (1943, 1962a, b) and has been discussed in detail by SKAPSKI, ORIANI, SEMENCHENKO, 1957a) (SIROTA (1943, 1962); TURNBULL and HOLLOMAN, and by TURNBULL 1950). We assume, as it is usually adopted that $\Delta Z = \Delta H\,(\Delta T/T_k)$. In this case the size of a three-dimensional spherical nucleus is $r_3 = 2V\beta\,(T_k/\Delta T)$ and the work on the formation of a three-dimensional nucleus of a new phase in the units of kT as a function of relative undercooling is (SIROTA 1948b, d)

$$A_3/(kT) = \gamma_3 \frac{\Delta H}{kT}\left(\frac{T_k}{\Delta T}\right)^2.$$

Similarly the work on formation of a three-dimensional nucleus as a function of relative change of any generalized force X is

$$A_3/(kT) \cong \gamma_{3i} \frac{\Delta H}{kT}\left(\frac{X_{ik}}{\Delta X_i}\right)^2.$$

B. Work on formation of two-dimensional nuclei as a function of relative undercolling and deviation of a generalized force from its equilibrium value

At analogous basis the work on formation of a two-dimensional nucleus as a function of relative undercooling supposing that the role of other factors is small, is

$$A_2/(kT) = \gamma_2 \frac{\Delta H}{kT} \frac{T_k}{\Delta T}.$$

The work on formation of two-dimensional nucleus as a function of relative deviation of any generalized force from equilibrium value is

$$A_2/(kT) = \gamma_{2i} \frac{\Delta H}{kT} \frac{X_{ik}}{\Delta X_i}.$$

Here γ_3, γ_{2i}, γ_2, γ_{2i} are proportionality quotients.

C. Work on formation of metastable-phase nuclei in case of undercooling

Analysis of the shapes of the plots of the work on formation of three- and two-dimensional nuclei of a stable phase 2 and metastable phases 2^a, 3^a, ... versus undercooling testifies that under adopted conditions the smallest work in the units of kT corresponds to the stablest phase 1 and the greatest work — to the least stable phase "a", if undercooling (or supersaturation) is small. At temperature lower than $1/3\ T_k$ for three-dimensional nuclei and lower than $1/3\ T_k$ for two-dimensional nuclei this relation reverses: the smallest work on the formation of three- and two-dimensional nuclei corresponds to the least stable phase "a".

Free energy curves for stable 1, 2 and metastable a_1, a_2, a_3 modifications of one-component system are shown in Figure 2.

The curves of the work on formation of three-dimensional (A_3/kT) and two-dimensional (A_2/kT) nuclei versus temperature are presented in Figure 3. In the process of computation the following values of the quantities have been assumed:

$$\Delta H_{21} = 1.43\ \text{kcal/mole}; \qquad T_{k1} = 1000\ \text{K}$$
$$\Delta H_{2a1} = 1.25\ \text{kcal/mole}; \qquad T_{a1} = 900\ \text{K}$$
$$\Delta H_{2a2} = 1.10\ \text{kcal/mole}; \qquad T_{a2} = 800\ \text{K}$$
$$\Delta H_{2a3} = 0.95\ \text{kcal/mole}; \qquad T_{a3} = 700\ \text{K}$$
$$U = 14.3\ \text{kcal/mole} \qquad \gamma_3 = 2.$$

As may be seen from the given Figure 3 when undercooling (or thermodynamic supersaturation) is small, most probable is appearance and growth of the stable phase nucleus. When supersaturation is great it is most probable that the least stable, furthest from equilibrium metastable phase will form and grow, since the work on its formation is the smallest.

It also follows from the carried-out analysis that Ostwald's step rule in its general form may be valid for one-component systems in case of sufficiently large supersaturation.

Elastic stresses arising as a result of the difference of specific volumes of initial and appearing phases or caused by imperfect conjugation of crystallochemical lattice of orginating phase with a base (or a seed) may to a certain extent change this relation in the intermediate domain of supersaturation, facilitating the formation of a phase with maximum crystallochemical affinity or crystallographic conformity with initial phase (or seed). However, the above conclusion remains valid in its general

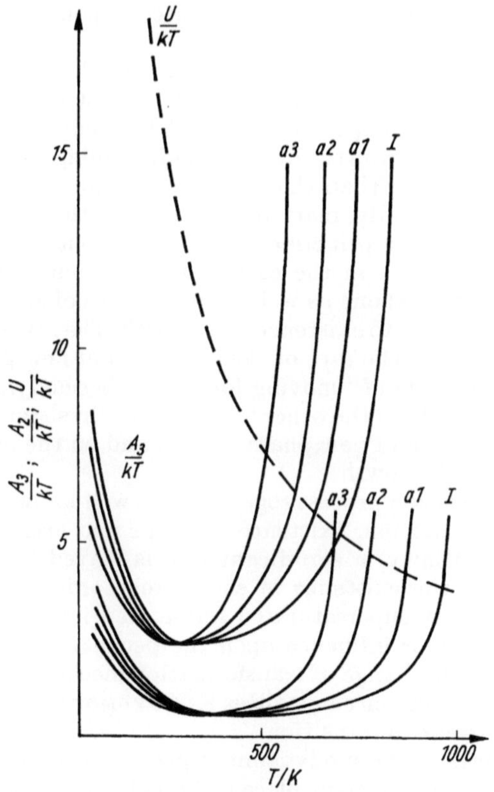

Fig. 3. Work on formation of three- and two-dimensional nuclei in the units of kT: A_3/kT and A_2/kT versus temperature for stable phase 1 and metastable phases a_1, a_2, a_3. Also shown is activation energy U/kT (U = 14.2 kcal/mole) Computation

form: in case of small thermodynamical supersaturation the smallest work on formation of nuclei corresponds to most stable phase and on the contrary in case of major thermodynamic supersaturation the smallest nucleation work corresponds to the farthest from equilibrium, least stable phase. In the range of intermediate supersaturations the substantial role is played by crystallochemical and crystallographic relations of appearing phases with initial phase or seed.

VII. Main factors causing polymorphic phase transition (crystallization)

Sufficiently complete physical substantiation of Ostwald's step rule and determination of its applicability limits is possible on the basis of detailed study of kinetics of the phase transformation (crystallization) of initial phase 2 into stable (under given conditions) phase 1 or metastable phases $a_1, a_2, \ldots, a_n$.

It is well-known (D. K. Chernov, 1868, 1898; Tammann 1903) that every phase transformation process takes place by means of formation of nuclei-"germs" of the new phase — the centres of crystallization and of their growth. Volmer-type process of growth via two-dimensional nucleation is not the only possible one. Other feasible processes are: the growth on dislocations, prominences; branchy (dendritic, needle-shaped), martensite growth, and also so-called "normal growth". The example of normal growth is the growth of fluid drops in supersaturated vapour: here there are no two-dimensional nuclei.

The difference of absolute values of linear rate of VOLMER growth via addition of 2-dimensional nuclei and that of normal growth in principle cannot be great, at least in the region of moderate undercooling (supersaturation). A low undercooling VOLMER growth rate exceeds the rate of normal growth. At great undercooling, conversely, the VOLMER growth rate may become lower than that of normal dislocational, dendritic growth, Widmanstette-type and naturally, martensite-type growth.

First order phase transitions are in principle fluctuational ones. Fluctuational character of transformation is revealed in the case of spontaneous occurance and growth of nuclei of new phase in initial one as well as in presence of seeds and bases where the germs of the new phase come to existence and growth. The rate of the transformation processes is determined by the rate of the formation of new phase nuclei, and depends finally on the magnitude of "moving force" — thermodynamic supersaturation, mobility of atoms — and, on the other hand, on the "resistance force" — the amount of work on the formation of new phase centres and on the movement of atomic layers in the process of their growth.

In crystallization or recrystallization process together with well-known factors (i.e. value of thermodynamic supersaturation, existence of shape and orientational correspondence, crystallochemical affinity) a significant role is played by various unaccounted factors stipulated by the existence of short-range order, "microheterogeneity", initial thermodynamically supersaturated phase, and its change with the growth of the supersaturation. Major influence upon the peculiarities of the transformation kinetics is exerted by the change of transformation mechanism and related change of activation energy. At similar circumstances KURDYUMOV (1930, 1949, 1952, 1960) analysed processes of martensite transition.

For example, with the increase of thermodynamic supersaturation (undercooling) three-dimensional nucleation sometimes gives place to branchy dendritic, Widmannstätten-type growth, trostomartensite and then to martensite-type transition mechanism. The shape of growing crystallites exerts substantial influence on the regularities of time development of the transformation process, including isothermal process, and on its temperature dependence.[6])

In undercooled fluids, like any other thermodynamically supersaturated phase, the formation of various quasi-stable "pre-nuclei" of new phase with different structures is taking place, their size and stability being comparable to, or even exceeding those of growing stable and metastable phase nuclei (SIROTA 1961).

In processes of phase transformation the role of suspension of such associations or complexes is rather considerable. Presence of this "plankton" (expression of A. A. BAJKOV) may modify or even determine the flow of transformation process, may cause preferential formation of a certain metastable, thermodynamically possible phase.

In a number of cases the formation of new-phase nuclei may take place via coalescence of pre-nuclei and their recrystallization. Formation of relatively stable short-range complexes and quasistable pre-nuclei in the initial phase may be conditioned by a number of processes. Among them are the following ones. In thermodynamically supersaturated transitional phases a new "quasi-phase" may sometimes originate from the closest packing of ions with space-type, instead of surface-type structure (Fig. 4). They may form via addition of ions to tetrahedric holes formed by three adjacent ions lying on the surface of the complex or pre-nucleus (SIROTA 1961a, 1967;

---

[6]) The dependence of the shape of growing crystals on undercooling had been analysed by ARTEMJEV, DAVENPORT and BAIN, MEHL and SMITH, FRIDLANDER (1948) et al.

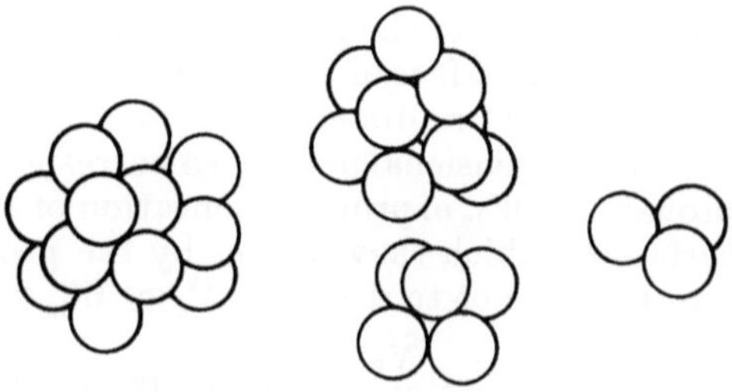

Fig. 4. Clusters forming via space dense packing containing 3, 5, 11, 17 atoms

SHTERENBERG). Existence of quasi-stable pre-nuclei of several critical sizes with various types of structure may be explained also by a certain dependence of surface energy on the size of new-phase pre-nuclei, by difference of specific volumes and correspondingly by difference of deformation field energy in the vicinity of appearing pre-nuclei, etc. Also processes of internal composition redistribution, internal adsorption (ARKHAROV), presence of Cottrell's atmosphere may exert major influence.

The role of these and other factors influencing the sequence of phase formation has not been cleared yet despite the significance of the question.

Later we shall confine ourselves to the qualitative consideration of metastable phase formation conditions in VOLMER-type processes — three-dimensional nucleation and subsequent growth of new phase nuclei within the initial phase. However, one must keep in mind that non-VOLMER-type processes described above may exert decisive influence upon the flow of transformation processes and the sequence of appearance of metastable phases (polymorphic modifications).

In order to analyse the succession and kinetics of the isolation of stable and metastable phases from thermodynamically supersaturated initial phase we shall briefly consider elementary processes of first-order phase transition, using classical ideas of GIBBS-VOLMER, VOLMER (1939), FARKASH, BECKER and DÖRING, STRANSKI and KAISCHEW, Ja. I. FRENKEL (1941, 1945), Ja. B. ZELDOVICH, SEMENCHENKO and PETROVSKII and the methods developed by the author (SIROTA 1945, 1948, 1957, 1962a, b), see also DANILOV; KUZNETSOV (1937, 1963), DUNNING.

The rate of formation of three-dimensional nuclei in the volume V of a mole of initial phase may be semi-quantitatively represented as a product of a number of factors: W_3'' — the probability of successive transformation into a nucleus of a new phase of an activated group of n_3 atoms having the critical volume; W_{3c} — the probability of concentration fluctuation in the result of which the composition of this group of atoms will correspond to the composition of appearing phase; P_3 — survival probability for the forming nucleus, N_D/n_3 — number of groups containing n_3 atoms each among N_D moving activated atoms in the volume V of a mole of initial phase in a unit time;

$$W_3 = \frac{N_D}{n_3} W_3'' \cdot W_{3C} P_3 ,$$

where[7])

$$N_D = N\nu e^{-U/kT} .$$

Here ν is the effective frequency of atom oscillations. The probability P_3 of survival of a three-dimensional nucleus will in its turn be determined by the product of the

[7]) TURNBULL and FISCHER 1949 (see also HOLLOMAN and TURNBULL 1953) using the method of absolute reaction rates, included kT/h into pre-exponential factor, instead of ν in present work, since it may be adopted that $h\nu = kT$. However it may be more convenient to use characteristic temperature θ substituting ν by $(k/h)\,\theta$.

probability W_2'' of the group of n_2 atoms, a progenitor of a two-dimensional nuclei, to occur on the surface of a three-dimensional nucleus, by the ratio F_3/F_2 of the surfaces of three- and two-dimensional nuclei, by the probability W_{2c} of the composition of this two-dimensional nucleus to correspond the composition of the new phase, by the probability $W_{2\eta}$ of proper orientation of atom or a group of atoms with respect to the surface to which they adjoin, by the probability P_2 for this nucleus to continue its growth, but not to dissolve, $P_2 \cong W_{02} \cong 1/2$. Thus,

$$W_3 \cong \frac{N_D}{n_3} W_3'' W_2'' W_{3c}' W_{2c} W_{2\eta} W_{02} (F_3/F_2) .$$

Similarly the probability of the appearance of a two-dimensional nucleus on a unit surface of a growing crystal in a unit time will be determined by a product of a probability W_2'' for the activated group of n_2 atoms to transform into a two-dimensional nucleus, by the probability W_{2c} of a concentration fluctuation in the volume of a two-dimensional nucleus from activated composition of initial phase to the composition of a new phase, by the probability $W_{2\eta}$ of proper orientation of the adding atom (or particle), by the number N_{2D}/n_2 of groups of n_2 atoms each among N_{2D} atoms, which is the difference of the number of atoms striking the unit surface of a growing crystal and the number of reflected atoms, by the probability P_2 for the nuclei not to dissolve

$$W_2 \cong (N_{2D}/n_2)\, W_2'' \cdot W_{2c} \cdot W_{2\eta} \cdot P_2 .$$

Linear growth rate c of a crystal to a first approximation may be adopted as $c \cong \cong W_2\delta$. (δ is thickness of two-dimensional nucleus.)

In accordance with Boltzmann's statistics it is assumed that

$$W_3'' \cong e^{-A_3/kT} , \qquad W_2'' \cong e^{-A_2/kT} ,$$

where A_3 and A_2 are respectively the works on the formation of three- and two-dimensional nuclei of critical volume.

Resulting functional dependence on ΔT of probabilities of appearance of three- and two-dimensional nuclei (concentration fluctuations are disregarded) is as follows:

$$W_3 \cong K'(\Delta T/T)^3\, e^{-K_1/T} \cdot e^{-K_2/T(T_k/\Delta T) - K_3/T(T_k/T)^2} \cdot e ,$$

$$W_2 \cong K''(\Delta T/T)^2\, e^{-K_1/T} \cdot e^{-K_2/T(T_k/\Delta T)} .$$

The work involved in the fluctuation of the nucleus composition leading from initial to new phase is in case of two-dimensional nuclei kTB_2n_2 and in case of three-dimensional nuclei kTB_3n_3 where

$$B_2 \cong (C - \alpha)^2/(2C_1\alpha_2) , \qquad B_2 \cong B_3 .$$

Using Gauss distribution we obtain the probability of the concentration fluctuation in the binary solution of phase 1 from active composition α_2 to C_1, corresponding to the composition of new phase;

$$W_{3c} \cong \eta_3\, e^{\frac{-(C_1-\alpha_2)^2 n_3}{2C_1\alpha_2}} ; \qquad W_{2c} \cong \eta_2\, e^{\frac{-(C_1-\alpha_2)^2 n_2}{2C_1\alpha_2}} ,$$

where

$$\eta \cong \frac{1}{\sqrt{2\pi C_1(1 - C_1)}} , \qquad \eta_2 \cong \eta_3 \cong \eta .$$

Calculation of concentration fluctuation probability can be compiled with the help of different methods: statistical thermodynamics, SMOLUCHOWSKI's statistical method which has been used by e.g. STARK et al. FINK and SMITH used Poisson distribution in computation of fluctuation probability. In our work we used Gauss distribution (see e.g. GNEDENKO, GLIVENKO).

Analysing the formation rate of the three-dimensional nuclei W_3, linear growth rate of nuclei and massive crystals $C \cong W_2\delta$ we must take into account the size of three-dimensional crystals, the growth rate of the two-dimensional nucleus along the growing face of a crystal, the dependence of surface energy on the size of chips and other factors, among them the size of the activity volumes in which the processes named take place in proper time.

It is assumed that the two-dimensional nucleus increases rapidly along the growing face of the crystal within the active region considered, the appearance of the new layer being caused by addition of one two-dimensional nucleus. It is assumed therefore that probability of formation of two-dimensional nucleus is related to the probability of origination of a new layer on the growing surface of the crystal. This may be true if growing crystals are relatively small and supercooling is also small.

In the analysis of the succession of formation (crystallization) of stable and metastable phases to the first approximation the dependence of the surface energy on the dimensions of the chips is not as a rule taken into account.

Let's remark that ultimately the linear growth rate C is determined by the difference of fluxes Φ_2 and Φ_1 of particles (atoms, ions or molecules) moving off the initial phase and off growing new phase, correspondingly, normally to the growing surface of the crystal, $C \cong K_d(\Phi_2 - \Phi_1) \cong Z_2\xi_2 - Z_1\xi_1$. The fluxes Φ_2 and Φ_1 are referred to a unit surface in a unit time; K_d is a coefficient including geometric and accomodation factors; ξ_1 and ξ_2 are kinetic coefficients depending on activation energy, structure of phase boundary and process mechanism. In thermodynamic equilibrium $\Phi_1 = \Phi_2$, $C = 0$. In case of thermodynamic supersaturation $\Phi_1 < \Phi_2$. The flux off initial to growing phase Φ_2 is in general greater than the flux Φ_1^0 conditioned only by its free energy per unit volume (Z_1/V): $\Phi_1 \cong \Phi_1^0\zeta, \zeta > 1$. This is due to peculiarities of surface structure caused by the transformation mechanism, to number of growth points, i.e. because of factors usually called "kinetic", which are in fact connected with local changes of surface energy.

Time evolution of the isothermal transformation of initial phase with sufficient accuracy is described by KOLMOGOROV-type expression see also KAZEYEV (1936, 1956), AVRAMI (1939), JOHANSON and MEHL, SIROTA (1942, 1950). This problem has been considered also by GÖLLER and SACHS; TAMMANN; LIPILIN; RODIGIN, etc., see ZHDANOV; BELENKII.

In the form given by SIROTA (1942, 1953) the expression describing kinetics of isothermal transformation process is as follows:

$$V_\tau = V_0\left\{1 - \exp\left[-\left(\frac{1-\nu}{2-\nu}\right)^{\frac{2-\nu}{1-\nu}} \cdot A^{\frac{1}{1-\nu}} \cdot \tau^{\frac{2-\nu}{1-\nu}}\right] + \dots + v_3^*\tau\right\},$$

where $A = K_sC_1$, $\nu = \frac{2}{3}$ for three-dimensional growth, $\nu = \frac{1}{3}$ for two-dimensional growth; $\nu = 0$ for one-dimensional growth; v_3^* is volume of a new phase centre which practically cannot grow, e.g. martensite "needle".

In the case of three-dimensional growth, according to KOLMOGOROV, by the moment τ the volume of transformed phase V_τ in the initial volume V_0 is $V_\tau = V_0(1 - e^{-1/3\pi W_3 C^3 \tau^4})$ KOLMOGOROV's formula and the analogous formula by JOHANSON and MEHL for finite volumes of initial phase are approximate though the main part of the process is described with good accuracy. If the linear growth rate if $C \cong W_2\delta$ and m is the fraction of the transformed phase, $m = V_\tau/V_0$, by the moment τ_m we have

$$\ln \tau_m = \tfrac{1}{4}\,\{\ln[\ln(1 - m)] - \ln \pi/3 - \ln W_3 - 3 \ln W_2 - 3 \ln \delta\}\,.$$

Earlier we remarked (SIROTA 1950) that these expressions were not exact, in particular for complete transformation time ($m \to 1$, $\tau \to \infty$) because in finite volumes which are parts of any initial volume and contain one or more growing nuclei appearing in initial finite time interval, the total transformation time cannot exceed $\tau_m = R/C_{min}$, where R is the radius of the sphere enclosing the volume of initial phase, and $C_{min} \cong$ $\cong (W_2\delta)_{min}$ the linear rate of isothermal growth of the nucleus, which appeared in the initial moment ($\tau = 0$).

Kinetics of isothermal processes of eutectic or eutectoid transformations, despite a series of specific peculiarities (presence of leading phase, special role of diffision; see BOCHVAR, BOCHVAR and TAMMAN, BOCHVAR and GOREV; CAHN) is formally described by the same equations.

VIII. Influence of thermodynamic supersaturation and of relative undercooling value upon the kinetics of formation of stable and metastable phases in one-component systems

Analysis of the given equation which explains experimental data comparatively well, permits to obtain a number of conclusions. Firstly, time τ_m of isothermal transformation of m-th fraction of initial volume, described by this equation, descreases with the growth of thermodynamic supersaturation at a given constant temperature.

Secondly, when supersaturated phase 2 transforms not only into stable phase 1 but also into metastable phases $a_1, \dots, a_n$, the time of transition into metastable phases shortens with the growth of phase 2 supersaturation with respect to each of them. Note that if $\Delta Z_i \to 0$ then for corresponding phase i $\ln \tau_{mi} \to \infty$.

Thirdly, if supersaturation (undercooling) is small, the stable phase has the smallest value of $\ln \tau_{mi}$. Conversely, if supersaturation (undercooling) is greater than a certain value ΔZ_μ the transition time will be the shortest for the least stable phase.

As it was remarked above, the surface energy σ_i on the boundary between initial and appearing phase i is approximately proportional to the energy of phase transition $\sigma_i \cong \beta \Delta H$. Consequently surface energy on the boundary initial phase — stable phase 1 is greater than on the boundaries initial phase — metastable phase, $\sigma_{21} >$ $> \sigma_{2a1} > \dots > \sigma_{2an}$, according to the relation $|\Delta H_{21}| > |\Delta H_{2a1}| > \dots > |\Delta H_{2an}|$.

Time of transformation of a certain m-th fraction of initial phase into stable phase 1 with increasing supersaturation ΔZ_{21}, decreases slower than that of metastable phases $a_1, \dots, a_n$. As a result of this there appear intersection points on the curves $\ln \tau_{m1}, \ln \tau_{ma1}, \dots, \ln \tau_{man}$.

In case of major supersaturation values, exceeding ΔZ_μ: $\Delta Z > \Delta Z_\mu$ the greatest rate and, consequently, the shortest time of transformation corresponds to the formation of the least stable phase a_n.

When temperature falls lower than T_k thermodynamic supersaturation grows and the work on formation of three- and two-dimensional nuclei A_3/kT, A_2/kT passes through minimum. At the same time the mobility of atoms which depends on activation energy U, decreases exponentially.

Consequently time logarithm $\ln \tau_m$ for isothermal transition of m-th fraction of initial phase has a minimum at a certain undercooling.

In particular, when thermodynamic supersaturation is due only to undercooling, $\Delta Z = \Delta H\,(\Delta T/T_k)$ then, as follows from GIBBS-VOLMER-FRENKEL' theory, the work on formation of three-dimensional nuclei A_3/kT has its minimum at $T = T_k/3$ and that for two-dimensional nuclei A_2/kT at $T = T_k/2$ and respectively time logarithm $\ln \tau_m$ will be practically near its minimum (which is given by the equation $\partial \ln \tau_m/\partial T = 0$).

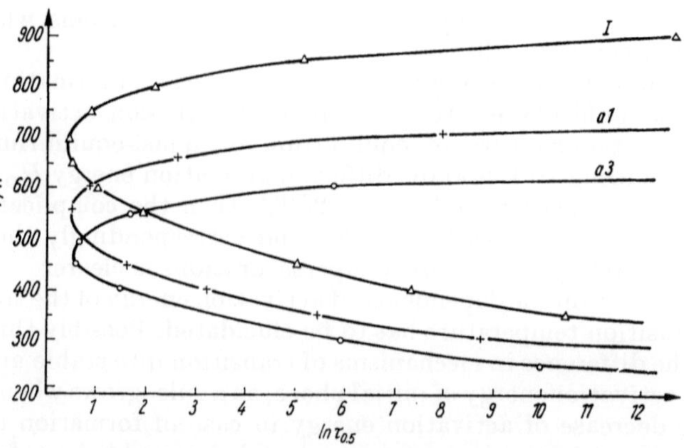

Fig. 5. Time logarithm for half-transformation of initial phase 2 into stable phase 1 (T_k = = 1000 K, U = 18 kcal/mole and into its metastable modifications (T_k = 800 K, U = 14.2 kcal/mole; T_k = 700 K, U = 12 kcal/mole). Computation $\tau(s)$

As can be seen from Figure 5, only for undercooling to the temperature lower than 500 K, the shortest transformation time $\tau_{0.5}$ ($m = 0.5$) corresponds to the least stable phase a_3, i.e. in the first instant the farthest from equilibrium phase will isolate in accordance with Ostwald's step rule. (See also Fig. 6.)

Consequently Ostwald's step rule in the case considered is valid only when supersaturation is large.

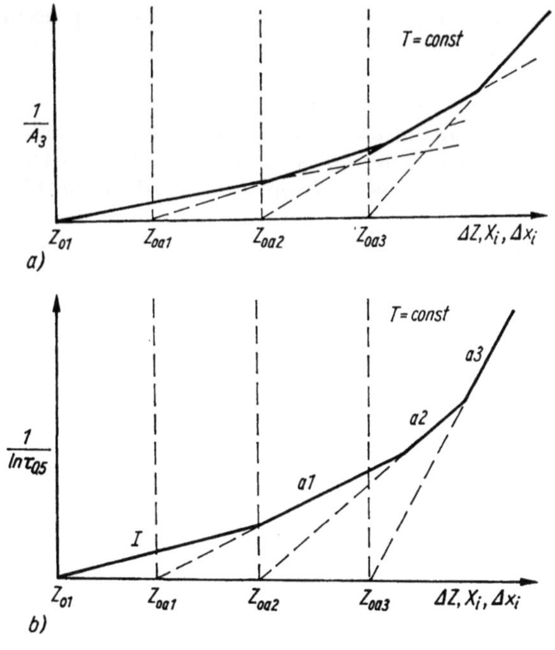

Fig. 6. Regions of thermodynamic supersaturation ΔZ, corresponding to a) the minimum nucleation work A_3 b) minimum time logarithm ln $\tau_{0.5}$ for half-transformation of initial phase 2 into stable 1 and metastable modifications a_1, a_2, a_3. A scheme

Here an important role is played by the diffusive mobility of atoms, which decreases exponentially as temperature falls.

Effective diffusion activation energy U is different in case of formation and growth of stable and metastable phases. If we adopt that diffusion activation energy is proportional to the temperature of equilibrium or quasi-equilibrium transition $U_i \cong aT_{ci}$ (e.g., according to Bugakov, diffusion activation energy U_D in crystals is related to the melting temperature T_s:$U_D \cong 20T_s$), then the complicated character of C-shape stability curves is clearly revealed and correspondingly the sequence of appearance of metastable phases at great supersaturations is clearer.

However, the question on the dependence of activation energy of the transformation process on the transition temperature has to be elucidated. Possibly this dependence is due mainly to the difference in mechanisms of transition into stable and metastable phases. Diffusion activationenergy of initial phase, as a rule, grows when temperature lowers. Therefore decrease of activation energy in case of formation of metastable phases with lower transition temperatures may result from additional factors, among them being: composition change of forming phases and, which is more important, difference in transition mechanisms (Kurdyumov), due to autocatalysis (Dehlinger 1939; Sirota 1955, 1967), elastic deformations caused by volume change in transformation (Nabarro, Cahn, Darken and Gurry) a.o.

The basis of diffusion theory of growth of a needle of new phase has been laid by Ivantsov (see also Hillert).

IX. Influence of the change of pressure and other generalized forces upon the sequence of formation of stable and metastable phases

When pressure is changed isothermally, with the growth of relative pressure deviation from the equilibrium value P_k, i.e. with the growth of thermodynamic supersaturation the transformation time logarithm decreases. If the formation of phases 1, $a_1, \ldots, a_n$ out of supersaturated with respect to them phase 2 is thermodynamically possible, when $p < p_a < p_{a1} < p_{21}$ the resulting curve $\ln \tau_m$ has a complicated character. If $p < p_\mu$ the smallest transformation time refers to the formation of the least stable phase a_n. When relative deviation of pressure from P_k is small, the stable phase is the first to isolate.

Similar regularities concerning succession of formation of stable and metastable phases out of thermodynamically supersaturated phase 2 in isothermal conditions take place when any of generalized forces X_i deviates from its equilibrium value X_{ik} causing the growth of thermodynamic supersaturation ΔZ (Fig. 6).

The dependence on ΔZ of work on formation of nuclei of stable and metastable phases and of half-transformation time (of transition of initial phase into stable and metastable phases) other things being equal will be determined by the surface energy σ_{2i} on the interfaces of initial phase with stable and metastable phases.

Influence of pressure and other generalized forces upon self-diffusion coefficient may change the shape of the curves $\ln \tau_m$ vs $\Delta p/p_k$ but practically does not change the above regularities with respect to succession of formation of stable and metastable phases, at least in the regions of small and great thermodynamic supersaturation.

Analysis of the results obtained shows that in case of small thermodynamic supersaturations resulting from deviation of pressure or any other generalized force from its equilibrium value, the most stable phase isolates in the first instance.

Conversely, in case of major thermodynamic supersaturations it is the least stable phase that appears the first. Consequently under considered conditions of pressure (or any other generalized force) change, Ostwald's step rule is valid also when deviation of pressure (or generalized force) from its equilibrium value is great.

As regards the sequence of formation of metastable phases a special role is played by the pressure change occuring in the transformation process because of the difference of specific volumes and elastic properties of initial and forming phases.

When new phase has a greater specific volume than the initial one, the pressure is growing causing slowing of the transformation process. A transition which takes place on the external surface of initial phase in vacuum and which is non-variant first-order phase transition, may adopt a character of polyvariant second-order transition within the volume of initial phase as a consequence of the pressure change. Significant hysteresis is exercised when the external factor (temperature, pressure) variation direction is reversed.

When thermodynamic supersaturation is sufficient, the probability of formation and growth of metastable phase nuclei with the smallest specific volume may be the greatest, and correspondingly, its separation will be preferencial, at least in the region of moderate thermodynamic supersaturation.

X. Influence of crystal bases and seeds upon the rate and succession of isolation of redundant stable and metastable phases

In presence of seeds and bases with sufficient dimensional and orientational correspondance and hence crystallochemical affinity with appearing phase (Dankov), the work on formation of a nucleus upon the base is approximately equal to the work on formation of two-dimensional nucleus and is in any case essentially smaller than that for three-dimensional nucleus in the absence of seeds.

In the case of complete crystallographic and crystallochemical correspondence transformation time logarithm for m-th fraction of initial phase is given by expression:

$$\ln \tau_m = B_m - \frac{U - \frac{4}{3} A_2 + \frac{1}{4} n_2^3 - B_2}{kT}.$$

Here we neglect fluctuations of concentration and deformation energy. As may be concluded from the above formula, the transformation time in the case considered is mainly determined by the work on formation of a two-dimensional nucleus. When composition of appearing phase differs from that of the initial one we must take into account the concentration fluctuation probability in its volume.

If the crystallochemical affinity of the base for one of the metastable phases takes place, nevertheless, if thermodynamic supersaturation is small or if formation of metastable phase is thermodynamically impossible, in the first instant only stable phase will form. In case of somewhat larger values of thermodynamic supersaturation the mentioned metastable phase, having crystallochemical affinity for the bases, will predominantly form upon the base.

However, even in the presence of bases at relatively large supersaturation most probably there will first appear the least stable phase out of all thermodynamically possible phases.

In Figure 7 are shown C-shape curves for the transition of phase 2 into phase 1 without bases and seeds and for transition of phase 2 into a metastable phase "a" in the presence of inclusions having crystallochemical affinity for the appearing metastable phase. As may be seen the resulting curve has a complicated shape, at temperature higher than T_μ stable phase predominantly forming, and at $T < T_\mu$ – metastable phase at, the most distant from equilibrium, forming. If crystallographic

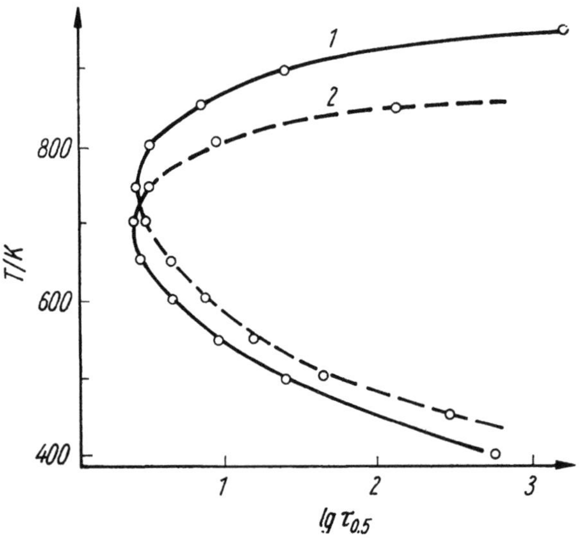

Fig. 7. Time logarithm for half-transformation of the phase 2 into stable modification 1 which has no crystallographic correspondence with initial phase and into metastable phase a which has such correspondence

correspondence of initial and appearing phases or base and appearing phases is incomplete the process of their formation on the boundary with the surface may depend on the degree of wetting of this boundary and on the change of surface energy. This effect was considered by MIKHNEVICH; FLETCHER; HOBBS; FUKS; ZAREMBA et al. a.o.

If degree of dispersion of the base is high, its structure may vary from that of a base of large crystallites (see SIROTA I). Hence the former may have crystallographic similarity (size and orientational correspondence, crystallochemical affinity) with the other, than the latter, metastable phase, stimulating its preferencial formation.

Substantial role may be played by nuclei having formed in canals, cavities, pores (KAZACHKOVSKII 1948, SIROTA 1962, TURNBULL 1950). This nuclei may be of both stable and metastable phase which depends on the structure and crystallochemical peculiarities of the base.

XI. Influence of composition difference of inital and forming phases

As regards the mechanism and kinetics of new phase formation out of thermodynamically supersaturated initial phase 2 a significant role is played by the probabilities of concentration fluctuation in the initial phase caused by composition difference, and by constituents of the work on formation of three- and two-dimensional nuclei depending on elastic properties, difference of specific volumes, degree of ordering of initial and appearing phases.

The influence of composition difference of initial and occuring phases may be illustrated as follows. As may be seen from the state diagram (Fig. 8), from the solution II of composition C_B of binary system A–B with decreasing temperature a stable phase 1 may isolate (with composition AB), together with its metastable modifications $a, \ldots, a_n$ and metastable phase $A_3B_2(\eta)$. The results of computation of temperature dependence of the work on formation of three- and two-dimensional nuclei

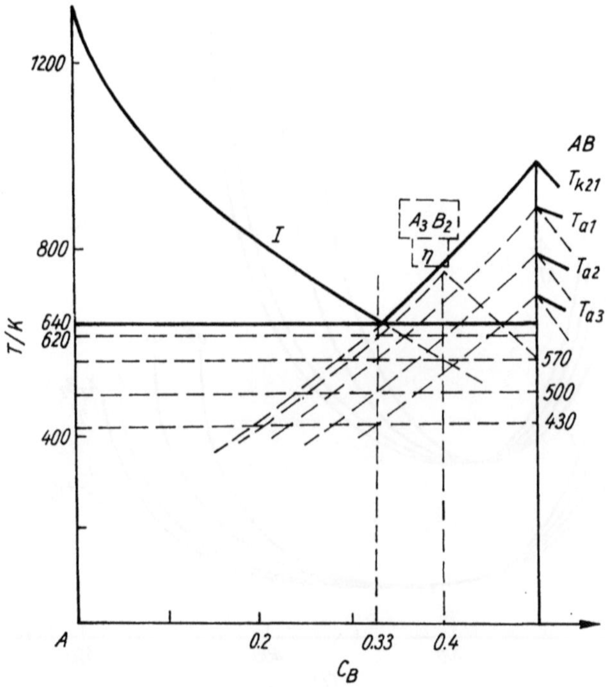

Fig. 8. State diagram: a stable compound AB (1) exists together with its metastable modifications a_1, a_2, a_3 and metastable compound A_3B_2 (η')

in the units of kT of all phases which can isolate from phase with $C_B = 0.33$, are shown in Figure 9. As may be seen from this figure the account of concentration fluctuation probability will result in the fact that excluding small supersaturations, when only stable phase forms, even in case of comparatively small or moderate supersaturation in the first instance metastable phase (η) is formed. The composition of the latter differs from that of initial phase less than from the composition of a stable phase AB. In case of large supersaturation, however, the least stable phase has the least total work on the nucleus formation; origination of this phase thus appears to be the most probable and transformation time involved is the shortest.

Computed curves A_3/kT, A_2/kT versus composition are shown in Figure 10 (concentration fluctuations are disregarded). The same curves with regard of concentration fluctuations, $A_3/kT + n_3B_3$, $A_2/kT + n_2B_2$ are shown in Figure 11. In Figure 12 there are presented half-transformation time logarithm curves for the transition of initial phase into stable and metastable phases computed for the state diagram of Figure 8.

Consideration of the role of concentration fluctuation probability, degree of undercooling and other kinetic factors permits us to ascertain why metastable state diagrams exist, e.g. Fe–Fe_3C (SIROTA) 1949; Fe–C (EPSTEIN; DARKEN, GURRY; ZHUKOV), Cd–Sb (SIROTA 1950a, b; MIROSHNICHENKO; PSARYOV).

Appearance of stresses E_g caused by elastic deformations arising from the volume change in the transformation process in this case of two- and multi-component systems may also significantly change the succession of formation of stable and metastable phases but mainly in the region of moderate supersaturation.

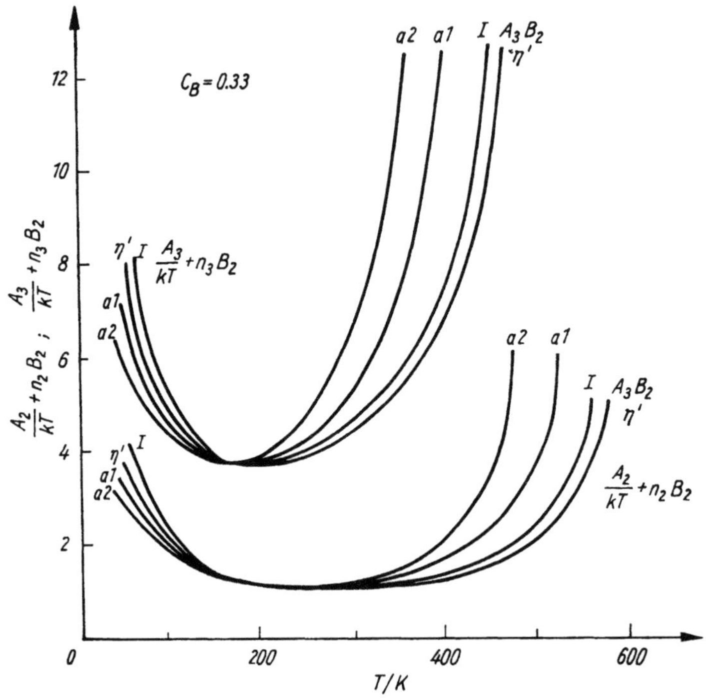

Fig. 9. Role of concentration fluctuation: temperature dependence of the work on formation of three-dimensional $\left(\frac{A_3}{kT} + n_3B_3\right)$ nuclei in the units of kT for stable modification 1 and metastable phases a_1, a_2, a_3 of compound AB, and metastable compound A_3B_2 (η′) Initial phase 2 (Z) has a composition $C_{BZ} = 0.33$

XII. Role of multiplicity of active faces of the base and orienting influence of external fields

It should be noted that multiplicity factor and influence of magnetic electric, gravitational, and other external fields upon thermodynamic supersaturation and hence upon the transformation process play not the least role. Crystallochemical affinity of occurring phase with a base, inclusions, seeds or with initial phase is possible only parallel to certain planes and along certain directions. Thus the effect of this inclusions or seeds on the transformation process, stimulating predominant formation of a certain stable or metastable phase having the closest crystallochemical affinity for the bases will not only depend on the degree of supersaturation, but also on the factor of multiplicity of analogous active faces and directions, on the probability of their proper coincidence, characterized by the above mentioned factor $W_{2\eta}$, and also on the effective value of conjugated active surfaces.

The role of external magnetic and electric fields must also be taken into account since they exert influence upon the value of thermodynamical supersaturation and the probability of coincidence of interacting planes and proper directions in the crystalline bases, the forming nuclei and new-phase crystals. Preferential directions imposed by the fields may also orientate solitary atoms, molecules and especially clusters adjoining to the growing crystals of new phase.

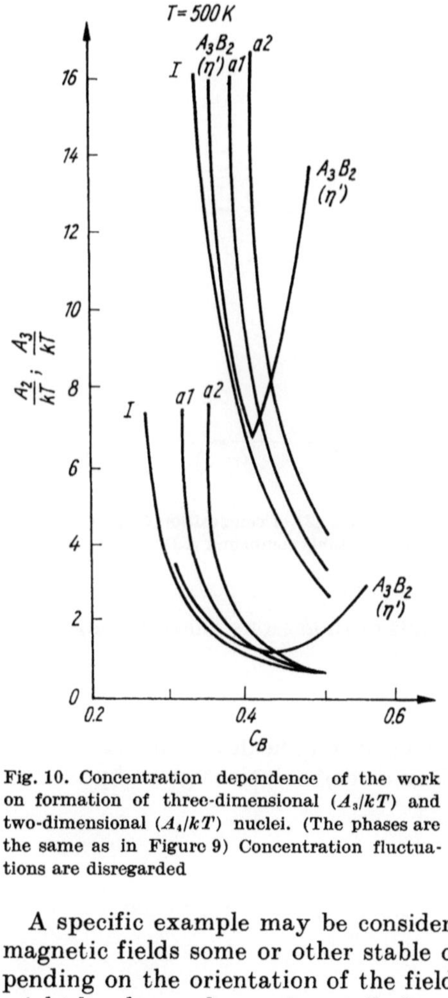

Fig. 10. Concentration dependence of the work on formation of three-dimensional (A_3/kT) and two-dimensional (A_4/kT) nuclei. (The phases are the same as in Figure 9) Concentration fluctuations are disregarded

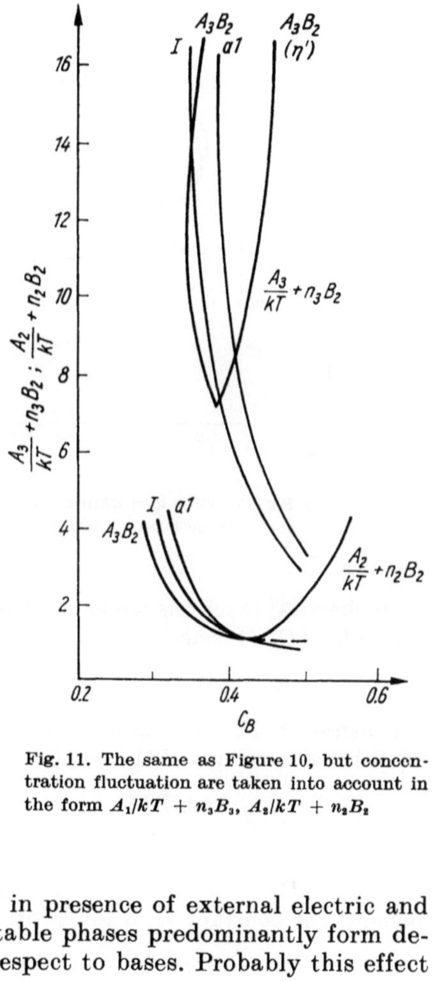

Fig. 11. The same as Figure 10, but concentration fluctuation are taken into account in the form $A_1/kT + n_3B_3$, $A_2/kT + n_2B_2$

A specific example may be considered when in presence of external electric and magnetic fields some or other stable or metastable phases predominantly form depending on the orientation of the fields with respect to bases. Probably this effect might be observed experimentally.[8])

XIII. Influence of thermodynamic supersaturation upon the formation of stable and metastable polymorphic modifications in chemical reactions, electrochemical processes and in process of film deposition

In chemical reactions, electrochemical processes, in case of condensation of vapour the value of thermodynamic supersaturation ΔZ may vary in wide range, and it is the value of ΔZ on which depends critical size of new-phase nuclei, work on their formation, rate of their formation and thus the rate of phase transition.

[8]) KAPITSA noted the role of orientating external fields on the example from archeology: determination of the position of pottery in furnaces according to its magnetization direction acquired in the Earth's magnetic field in the process of baking.

An example of similar influence is magnetization of marsh ores in the direction N–S in the process of their deposition in the terrestrial magnetic field.

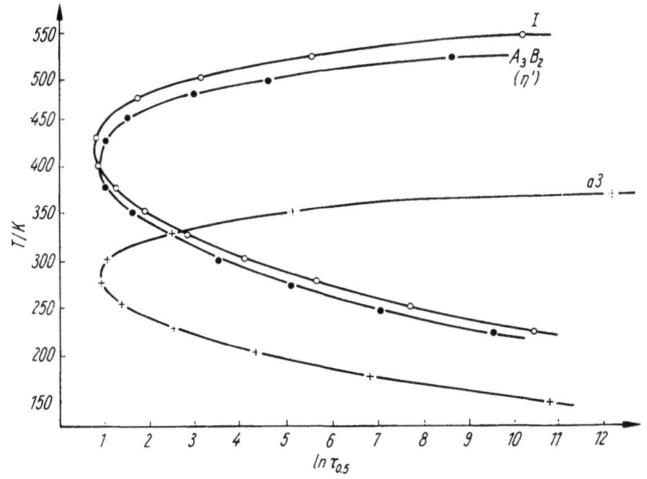

Fig. 12. Time logarithm of half-transformation of phase 2 (Z) of composition $C_{BZ} = 0.33$ into stable phase 1 and metastable modification a_3 and metastable compound A_3B_2 (η')

In chemical reactions which take place in liquids or in gaseous phase and in which solid phase C is formed,

$$A + mB \rightleftarrows C + nD$$

the degree of thermodynamic supersaturation depends on the deviation $K_2 - K_1$ of equilibrium constant of the reaction K_1 which is in case of solutions given by

$$K_1 = \frac{[A]\,[B]^m}{[D]^n},$$

where [A], [B], [D] are concentrations of the components. In case of gaseous components the constant K_1 is given by

$$K_1 = \frac{p^A p_B^m}{p_D^n},$$

where p_A, p_B, p_D are vapour pressures. The supersaturation ΔZ is then

$$\Delta Z = RT \ln K_2/K_1 .$$

In electrochemical processes the thermodynamic supersaturation ΔZ is determined by the deviation of potential difference between electrodes from its equilibrium value U_1:

$$\Delta Z = P(U_2 - U_1)$$

where F is Faraday constant.

The above consideration of the influence of thermodynamic supersaturation upon formation of stable and metastable phases and polymorphic modifications is valid in the case of chemical reactions, electrochemical processes, and also in the processes of crystal film deposition with the help of molecular beams (in the latter case $\Delta Z = RT \ln p_2/p_1$ where p_1 is equilibrium pressure of vapour of a given compound above the film, p_2 is the real pressure).

In the considered cases of chemical reactions and electrochemical processes formation of metastable phases in accordance with Ostwald's step rule will also take place when thermodynamic saturation is sufficiently high.

Important questions in electric crystallization had been studied by VOLMER (1922, 1934), GORBUNOVA (1935, 1957), GORBUNOVA and DANKOV (1949a, b), GORBUNOVA, IVANOVA and ZHUKOVA (1949), VAGRAMIAN (1950), KAISCHEW and BLIZNAKOV (1948), KAISCHEW, BUDEVSKII and MALINOVSKII (1955) and others.

An interesting question is the study of growth of electrolytic thin filaments. It was shown that the rate of their growth in isothermal conditions in a given electrolyte is determined mainly by potential difference and does not depend practically on the strength of current which influences only the thickness of the filament.

When thermodynamic supersaturation conditioned by concentration change of the components, their vapour pressure and temperature in chemical reactions, and also by the change of voltage in electrochemical processes, exceeds certain values, there appears possibility of formation and growth, in the first instance, of metastable polymorphic modifications, intermediate phases and states.

The dependence of work on formation of two- and three-dimensional nuclei on the degree of thermodynamic supersaturation, shown in Figure 6, is applicable in the cases considered above.

XIV. General regularities of the succession of formation of stable and metastable phases and limits of Ostwald's step rule validity

In all the cases considered a general regularity is preserved. If supersaturation or undercooling is small only stable phase is thermodynamically possible to form.

In a comparatively wide interval of intermediate values of thermodynamic supersaturation the transformation process and the type of the phase predominantly formed to a major extent is determined by supplementary kinetic factors: crystallographic correspondence of appearing and initial phase, presence of bases, having crystallochemical affinity with stable or some allowed metastable phase, concentration fluctuation probability, multiplicity factors of active faces of bases and forming phases, orientation by external fields, formation of nuclei in canals and pores of the base, etc.

In case of large values of thermodynamic supersaturation (e.g. undercooling), excluding may be the situations with great difference of specific volumes and/or composition of initial and appearing phases, in the first instant the farthest from equilibrium phases form. Among these metastable, non-equilibrium phases are those which can exist in equilibrium only under high pressure or in strong magnetic and electric fields.

Thus the famous Ostwald's step rule in the most general form holds only in case of sufficiently large thermodynamic supersaturations.

The above considerations allow with a sufficient certainty to reveal different physical factors causing the formation of metastable phases and to establish the conditions of Ostwald's step rule validity and failure.

XV. Several examples of formation of metastable polymorphics modifications and other phases in the process of crystallisation and phase transition

The ideas developed here are not only of physical interest but undoubtedly have a major practical significance. A considerable amount of available experimental results proves the proposition presented in the article. Among them is, for example, deposition of thin films of cubic boron nitride and diamond in situation where graphitic modifications are thermodynamically stabler.

The formation of metastable phases results in complicated types of C-shaped curves of undercooled austenite in steels, supercooled solid solutions in alloy on the basis of titanium and zirconium, of dispersional solidification of aluminium alloys (reference book by NOVIKOV; FRIDLANDER et al.).

Stability of undercooled austenite of the majority of complexly alloyed steels containing carbide-forming elements (in contrast to carbon steel) is as a rule described by complicated C-shape curves (DAVENPORT, BAIN).

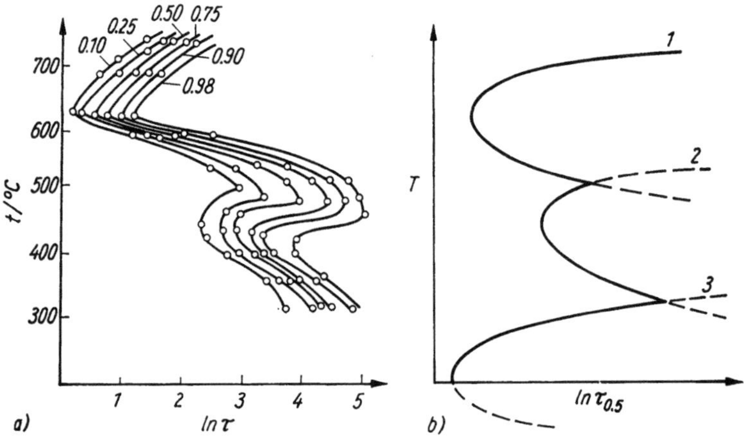

Fig. 13. a) Transformation of undercooled austenite of steel (1 % C, 1.5 % Cr): time logarithm for transformation of 10, 25, 50, 75, 90, and 98 % of mass; b) Schematic curves of time logarithm for half-transformation of austenite of chromium high frost-resistent steel. 1 — isolation of carbide $(Fe_2Cr)_7C_3$; 2 — isolation of carbide $(FeCr)_3C$; 3 — formation of trostomartensite

The term "complicated type" of C-curves of stability has been introduced by SADOVSKII and was primarily applied to steels; now its meaning has widened. An example of complicated-type C-curve of stability of initial undercooled phase is that of undercooled austenite of high-carbon chromium bearing steel (1.5% Cr, 1% C), as shown in Figure 13a (SIROTA, 1962). Also shown schematically (Fig. 13b) are simple stability curves of undercooled austenite with respect to the isolation of stable and metastable phases (modifications) of appearing carbides. This figure explains, how complicated C-curve forms, being the result of simple C-curves for various modifications. Comparison of the experimental complicated C-curve with computed ones allows one to clear up reasons and underline factors conditioning the complication of C-curves.

Let's remark that isolation of free carbon out of undercooled austenite as well as in the case of cementite graphitization for sufficiently large supersaturations, is possible in the form of its metastable polymorphic modifications, including diamond-like.

Similar conditions determine possibility of formation of metastable phases in case of thin film evaporation upon sufficiently undercooled bases (e.g. during deposition of carbon, boron nitride and compounds $A^{III}B^{V}$ there may form, respectively, diamond, cubic boron nitride and metallic films.

Successiveisolation of stable and metastable phases — Guinier-Preston's zones — is realized in the ageing of duralumin (see e.g. review by BUJNOV and ZAKHAROVA, KOMAR, BUJNOV, monograph by ZAKHAROVA).

In case of relatively small undercooling stable Θ phase is isolated ($CuAl_2$). At greater undercooling metastable Θ' phase forms, and at still lower temperature Θ'' phase forms. When ageing is completed (i.e. metastable phase has been isothermally isolated), subsequent heating to the temperature lower than equilibrium miscibility level but higher than that for metastable phase leads to dissolution of metastable phase. Thus may be explained the recovery phenomenon — GAILER's effect (GAILER; SIROTA 1945a). Further soaking at the pointed higher temperature leads to isolation of stable Θ phase.

The example of complicated-type C-shaped curves for light aluminium alloys, etc. may be found in reference book by NOVIKOV.

The above consideration explains the peculiarities of composition change of carbides during tempering of hardened, e.g. chromium steels, and in the isothermal transformation of undercooled austenite of these steels. At low temperatures (i.e. great thermodynamic supersaturations) predominantly metastable carbide phase is formed.

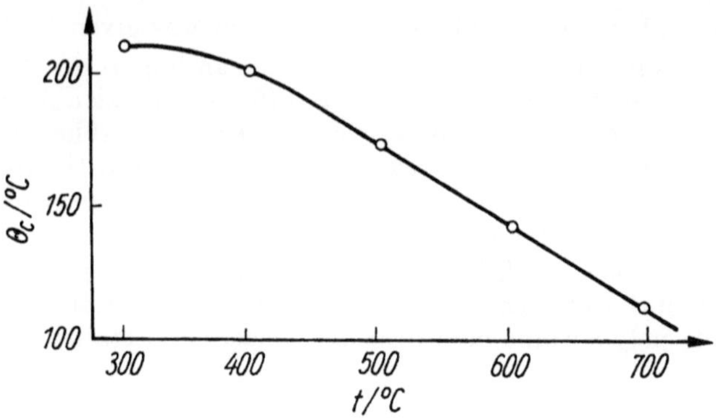

Fig. 14. Change of Curie point θ of the isolating carbides as a function of transformation temperature t (°C) (tempering of hardened steel 1 % Cr; 1 % C)

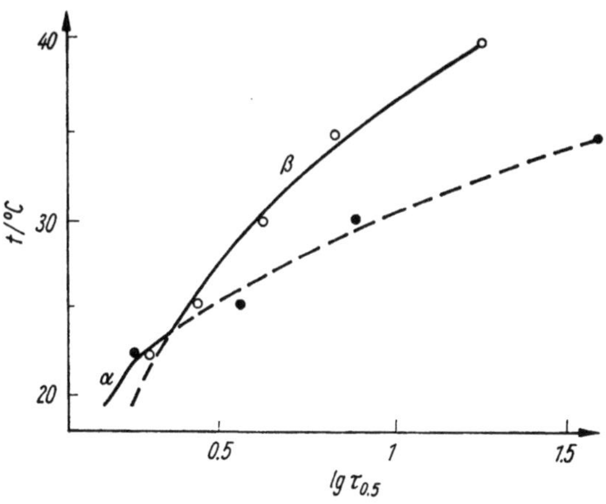

Fig. 15. Time logarithm of half-transformation of amorphous trilaurine into stable β and metastable α phases as a function of temperature (computed with the data of RAVITCH et al.)

In a number of our works the composition of carbides, isolating in isothermal transformation of austenite and during tempering of hardened steels had been controlled by the change of Curie point (magnetic transition temperature) (SIRCTA 1939, 1948) (Fig. 14). The observed regularities of carbide composition and structure change had been explained on the basis of similar ideas.

Persuasive example proving the views presented above is the formation of stable modification at relatively small undercooling and metastable α-modification at large undercooling in the process of isothermal crystallization of trilaurine, investigated by RAVICH et al. (Fig. 15).

Study of crystallization of thin amorphous films of stibium (SHALNIKOV), germanium (SIROTA, DENIS), aluminium oxide (SIROTA, SHOKHINA) proves that at sufficiently low temperatures metastable phases or metastable polymorphic modifications crystallize and, conversely, at higher temperatures, stable modifications of redundant phase form in the first instance.

Investigation of crystallization of thin $A^{III}B^{V}$-compound films shows that in case of great undercoolings metastable phases form including metallic superconductive phases (see collected articles edited by ALEKSANDROV and PETROSIAN; PALATNIK et al.; PALATNIK).

An interesting example of metastable phase formation was given by R. W. NERZ.

Formulation by Ostwald of his step rule has become an important stimulus in the study of formation of metastable phases and different metastable polymorphic modifications. Investigations presented above allowed to establish the limits of Ostwald's step rule validity, to find new regularities, to determine the conditions and succession of formation of stable and metastable phases. All these opens wide prospects of practical application of the established regularities.

The purpose of the present work has been further development of this important problem — experimental and theoretical analysis of the conditions leading to formation of metastable polymorphic modifications, phases and states.

References

ALEKSANDROV, A. P.: Nauka strane, Moscow 1983

ALEKSANDROV, L. N.: Kinetika obrasovanija i struktury tverdykh sloyov, Novosibirsk 1972

ALEKSANDROV, L. N., PETROSIAN, V. I.: Plenki dla mikroelektroniki, Novosibirsk 1977

ANOSOV, W. Ja.: Izv. sekt. Fiz-chim. analiza **22**, 67, 1953

ANOSOV, W. Ja., POGODIN, S. A.: Osnovnyye nachala fiziko-khimicheskogo analiza, Moscow 1947

ANOSOV, V. Ja., OZEROVA, M. U., FIALKOV, Yu. A.: Osnovy fiziko-khimicheskogo analiza, Moscow 1973

ARKHAROV, V. I.: Okisleniye metallov pri vysokikh temperaturakh, Sverdlovsk-Moscow 1945

ARKHAROV, V. I.: Fiz. Met. i Metalloved **12**, No 6 (1961)

ARTEMJEV, V. N.: Metod kristallizatii Sharov. Tr. Petr. Obchestva Estestvoispitatelei **37**, 59 (1914)

AVRAANI, J.: J. Chem. Phys. **7**, 103 (1939); **8**, 212 (1940)

BAJKOV, A. A.: Sobraniye trudov **1—5**, Moscow-Leningrad 1948—1952

BALARIN, M., SCHENK, M.: phys. stat. sol. **17**, 91 (1966)

BARABOSKHIN, A. N., KOSIKHIN, L. T.: Kristallizatsiya i fazovyye prevrascheniya, Minsk 1971, pp. 118—125

BARTENEV, G. M.: Zh. Fiz. Chim. **34**, 3 (1960)
BAZAROV, J. P.: Termodinamika, Moscow 1976
BECKER, R.: Z. Metallkde. **29**, 245 (1937); **32**, 128 (1938)
BECKER, R., DÖRING, W.: Ann. Phys. **24**, 719 (1935)
BELENKII, V. Z.: Geometriko-veroyatnostnyye modeli kristallizatsii, Moscow 1980
BELOV, A. F., GLAZOV, V. H., POTEMKIN, A. Ya., AKOPIAN, R. A.: Doklady AN SSSR **259**, 135 (1981)
BELOV, K. P.: Magnitnyie Prevrashcheniya, Moscow 1959
BELOV, B. V.: Struktura ionnykh kristallov metallicheskikh faz, Moscow 1947
BERTAUT, F. J.: Phys. Rad. **12**, 252 (1951)
BLANTER, M. Ye.: Zh. Tekhn. Fiz. **17**, 1731 (1947)
BLIZNAKOV, G. M.: Bulg. AN, ser. fiz. **6**, 301 (1951)
BLIZNAKOV, G. M.: Kristallografiya **4**, No 2 (1959)
BLIZNAKOV, G. M.: Godishnik na Sof. Univ. Phys.-Mat. **2**, 69 (1956)
BOCHVAR, A. A.: Issledovaniye mekhanizma i kinetiki kristallizatsii splavov elektricheskogo tipa, Moscow 1935
BOCHVAR, A. A., GOREV, K. V.: Izv. sekt. fiz.-khim. analiza, **8**, 283 (1936)
BOCHVAR, A. A., TAMMAN, G. A.: Vestnik Metallopromyshlennosti N 5–6 (1927)
BOKII, G. B.: Vvedeniye v kristallokhimiyu, Moscow 1960
BORELIUS, G.: Ann. Phys. **33**, 517 (1938)
BORELIUS, G.: Ann. Phys. **28**, 507 (1937)
BORISOV, V. T., DUKHIN, A. I.: in: Mekhanism i kinetika kristallizatsii Minsk 1969, pp. 169–181
BROWN, P. J., FORSYTH, J. B.: The crystal structure of solids, London 1973
BROWN, P. J.: Acta Cryst. **10**, 133 (1967)
BUGAKOV, V. Z.: Diffusiya v metallakh i splavakh, Moscow-Leningrad 1949
BURTON, W. K., CABRERA, N., FRANK, F. C.: Phys. Trans. **243**, 299 (1951)
BUYNOV, N. N., ZHURAVLEV, M., KOMAR, A. P., CHUFAROV, G. I.: Doklady AN SSSR **22**, 27 (1939)
BUYNOV, N. N., ZAKHAROVA, R. R.: Raspad metallicheskikh tverdykh rastvorov, Moscow 1964
BYKHOVSKII, A. I., LARIKOV, L. N., OVSIYENKO, D. E.: Kristallografiya **6**, 284 (1961)
CAHN, J. N.: Acta Metallurg. **9**, 795 (1961); **10**, 179 (1962)
CHATELIER, H., le: Compt. Rend. **9**, 118, 698 (1894)
CHERNOV, A. A.: Uspekhi Fiz. Nauk **73**, No 2 (1961)
CHERNOV, D. K.: Zapiski rus. tehn. obshchestva No 7, 339 (1868)
CHERNOV, D. K.: Staleliteinoye delo, St. Petersburg 1898
CHERNOV, D. K.: Izbrannye trudy, ed. by V. D. SADOVSKII, compiled by A. S. FEDOROV, Moscow 1983
CHRISTIAN, J. W.: Transformations in metals and alloys, second ed., pt. I, Oxford 1975
DANILOV, V. I.: Stroyeniey i kristallozatsiya zhidkosti, Kiev 1956
DANILOV, V. I., KAMENETSKAYA, D. S.: Zh. Fiz. Khim. **XXII**, 69, (1948)
DANILOV, V. I., KAZACHKOVSKII, O. D., LAPKOVSKII, I. M.: Zh. Eksp. Teor. Fiz. 18, No 10, 886 (1948)
DANILOV, V. I., RADCHENKO, I. V.: Zh. Eksp. Teor. Fiz. **7**, 1153 (1937)
DANKOV, P. D.: Doklady AN SSSR **23**, 548 (1938)
DADKOV, P. D.: Izv. Sekt. Fiz. Khim. analiza **16**, 82 (1941)
DARKEN, L. S., GURRY, R. W.: Physical Chemistry of Metals, New York 1953
DAVENPORT, E. S., BAIN, E. C.: Trans. AIME **90**, 117 (1930)
DEHLINGER, U.: Chemische Physik der Metalle u. Legierungen, Leipzig 1939
DEHLINGER, U., BORELIUS, G.: Handbuch der Metallphysik, Bd. I. Leipzig 1939
DUNNING, W. J.: in: Chemistry of the solid state, ed. W. E. GARNER, London 1955
EHRENFEST, P.: Communication Leiden, Suppl. 756, 1933
EPSTEIN, P.: The Alloys of Iron and Carbon, New York, 1936
FARKASH, L.: Z. Phys. Chem. **125**, 236 (1927)
FEDOROV, A. S.: Tvortsy nauki o metalle, Moscow 1980
FINK, W., SMITH, D.: Met. Techn. No 3, 95 (1939)

FLETCHER, N. H.: J. Chem. Phys. **31**, 1136 (1959)
FRANK, F. C., VAN DER MERVE, J. H.: Proc. Roy. Soc. **198**, 216 (1949)
FRANK, F. C.: Acta Metallurg. **1**, 15 (1953)
FRANK, F. C., READ, W. T.: Phys. Rev. **79**, 722 (1950)
FRENKEL', Ya. I.: Izv. Sect. Fiz.-Khim. Analiza **16**, 50 (1941)
FRENKEL', Ya. I.: J. Chem. Phys. **1**, 200, 538 (1939)
FRENKEL', Ya. I.: Zh. Ekop. Teor. Fiz. **16**, 1 (1948)
FRENKEL', Ya. I.: Kineticheskaya teoriya zhidkosti, Leningrad 1945
FRIDLANDER, I. N., STEPANOVA, M. G., SEFUKOV, O. A., BRUKHOVETZ, A. A.: Aluminiyevye splavy, Moscow 1968, pp. 226–230
FRIDLANDER, I. N.: Issledovaniye form rosta kristallov v zavisimosti ot skorotsti okhlazhdeniya, Moscow 1948
FRIDLANDER, I. N.: Aluminievye deformiruemye konstruktsionnye splavy, Moscow 1979
FUKS, N. A.: Uspekhi Fiz. Nauk **15**, 496 (1935)
GAYLER, M. L. V.: J. Inst. Metals **29**, 212 (1922)
GIBBS, J. W.: Thermodynamische Studien, aus dem Englischen übersetzt von W. OSTWALD, Leipzig 1882
GIBBS, J. W.: Termodinamicheskiye raboty, V. K. SEMENCHENKO, ed. Moscow 1950
GINZBURG, V. L.: Zh. Eksp. Teor. Fiz. **17**, 838 (1947)
GLIVENKO, V. I.: Kurs teorii veroyatnostey, 1939
GNEDENKO, B. V.: Kurs teorii veroyatnostey, Moscow 1961
GÖLLER, F., SACHS, G.: Z. Physik **77**, 281 (1932)
GORBUNOVA, K. M.: Izv. AN SSSR No 5–6, 1175 (1935)
GORBUNOVA, K. M.: Sbornik Rost Kristallov **1**, 1957
GORBUNOVA, K. M., DANKOV, P. D.: Uspekhi Kkumii **17**, 710 (1948)
GORBUNOVA, K. M., DANKOV, P. D.: Zh. Fiz. Khim. **23**, 616 (1949)
GORBUNOVA, K. M., ZHUKOVA, A. I.: Zh. Fiz. Khim. **23**, 605 (1949)
GORSKII, F. K.: Zh. Eksp. Teor. Fiz. **6**, 278 (1936)
GUGGENHEIM, E. A.: Mixtures, Oxford 1952
GUNIER, A.: Compt. Rend. **206**, 1641 (1938)
HARDY, H. K., HEAL, T. Y.: The Mechanism of Phase Transformation in Metals, London 1956
HARDY, H.: J. Inst. of Metals **77**, 5 (1950)
HOBBS, P. V.: Ice Physics, Oxford 1974
HOLLOMAN, Y. H., TURNBULL, D.: Progress in Metal Physics **4**, 333 (1953)
HUME-ROTHERY, W., REYNOR, G. V.: The Structure of Metals and Alloys, London 1962
IVANOV, O. S.: Izv. AN SSSR OKhN. No 4, 192 (1944)
IVANTSOV, G. P.: DAN SSSR **58**, 567 (1947)
JACOBS, P., TOMPKINS, F.: Chemistry of the Solid State, London 1955
JOHANSON, W. A., MEHL, R. F.: Tr. AIME **135**, 416 (1939)
KAISIHEW, R., BLIZNAKOV, G.: Compt. Rend. Acad. Bulg. **1**, No 2–3, 23 (1948)
KAISCHEW, R., BUDEVSKI, E., MALINOVSKI, J.: Z. Phys. Chem. **204**, 348 (1955)
KAISCHEW, R. A., STRANSKI, I. N.: Z. Phys. Chem. **170**, 295 (1934)
KAPITSA, P. L.: Private communication
KAZACHKOVSKII, O. D.: Sbornik nauchnykh rabot laborat. metallofiziki, Kiev 1948, p. 76
KAUFMAN, L., BERNSTEIN, H.: Computer calculation of phase diagrams, New York 1970
KAZEEV, S. A.: Metallurg. **6**, 43 (1936)
KAZEEV, S. A.: Kinetika v prilozheniyakh k metallovedeniyu, Moscow 1956
KHACHATURIAN, A. G.: Teoriya Fazovykh prevraschenii i Struktura tverdykh rastvorov, Moscow 1974
KHAMSKII, E. V.: Kristallizatsiya iz rastvorov, Leningrad 1967
KIRKWOOD, J. G., BUFF, J.: J. Chem. Phys. **17**, 338 (1949)
KLEBER, W.: Einführung in die Kristallographie, Berlin 1965
KOLMOGOROV, A. N.: Izv. AN SSSR, Mat. No 3, 365 (1937)
KOMAR, A. P.: Izv. AN SSSR Fiz. **11**, 497 (1947)
KREBS, H.: Grundzüge der anorganischen Kristallchemie, Stuttgart 1968

Kubaschewski, O.: Phase Stability in Metals and Alloys, New York etc. 1967.
Kurdyumov, G. V.: Rentgenografiya i primenenie rentgenovskikh luchey, Moscow 1936
Kurdyumov, G. V.: Zh. Tekhn. Fiz. 18, 999 (1949)
Kurdyumov, G. V., Sachs, G.: Vestnik metallo-promyshlennosti No 9—10, 165 (1930)
Kurdyumov, G. V., Sachs, G.: Z. Physik 64, 325 (1930)
Kurdyumov, G. V.: K teorii martensitnogo prevrashcheniya Problemy metallovedeniya i faz metallov, No 3, Moscow 1952
Kurnakov, N. S.: Vvedeniye v Fit. Khim. analiz, Moscow 1941
Kuznetsov, V. D.: Fizika tverdogo tela 1, Tomsk 1937
Laar, L. J. van: Die Thermodynamik einheitlicher Stoffe und binärer Gemische, Gröningen 1936
Landau, L. D.: Zh. Eksp. Teor. Fit. 19, 137, 627 (1937)
Landau, L. D., Lifshits, E. M.: Statisticheskaya Fizika, pt 1, Moscow 1976
Lerinman, R. M., Buinof, N. N.: FMM 10, 374 (1960)
Lemlein, G. G., Dukova, E. D.: DAN SSSR 102, 77 (1955)
Lemlein, G. G.: Vestnik AN SSSR, 4, 119 (1945)
Leontovich, M. A.: Vvedeniye v termodinamiku, Moscow 1951
Lipilin, I. P.: Kachestvennaya Stal., No 9, 1935
Lumsden, J.: Thermodynamics of Alloys, London 1952
Makarov, E. S.: Stroyeniye tverdykh faz s peremennym chislom atomov v elementarnoy yacheike, Moscow 1947
Maksimova, O. P.: Problemy metallovedeniya i fizika metallov, Moscow 1952
Mehl, R. F., Smith, D.: Metals Technology, No 521 (1934)
Mezhidov, V. I., Gurov, A. I., Krasnobrizhev, V. G.: Zh. Fiz. Khim 52, 52, 1965 (1978)
Mikhnevich, G. L.: Zh. Eksp. Teor. Fiz. 9, 215 (1939)
Mikhnevich, G. L.: Kolloidny Zhurn. 21, No 1, No 3, No 4 (1959)
Miroshnichenko, I. S.: Zakalka iz zhidkogo sostoyaniya, Moscow 1982
Mlodzeyevskii, A. B.: Termodinamika i teoriya faz Moscow-Leningrad 1922
Mlodzeyevskii, A. B.: Geometricheskaya termodinamika, Moscow 1956
Mlodzeyevskii, A. B.: Teoriya faz, Moscow 1937
Mott, N. F., Nabarro, F. R. N.: Proc. Roy. Soc. 52, 86 (1940)
Mýl, I.: in: Rost Kristallov 5, 246 (1965)
Nabarro, F. R. N.: Proc. Phys. Soc. 52, 90 (1940); Proc. Roy. Soc. A. 175, 519 (1940)
Nerz, R. W.: Tr. Farad. Soc 5 (1949)
Nikitina, G. V., Romanenko, V. N.: Izv. AN SSSR, No 6,156 (1964)
Novikov, I. I., ed.: Diagrammy isotermicheskogo raspada rastvora v aluminiyevych splavakh, Moscow 1973
Nývlt, T.: Collect. Czech. Chem. Comm. No 28, 2269—2279 (1963)
Oriani, R. A.: J. Chem. Phys. 18, 575 (1950)
Ostwald, W.: Grundriß der Allgemeinen Chemie, Leipzig 1899
Ovsiyenko, D. Ye.: Problemy sovremennoy kristallografii, Kiev 1975, pp. 127—149
Palatnik, L. S.: Doklady AN SSSR, 82, No 6 (1952)
Palatnik, L. S., Fuks, M. I., Kosevich, V. M.: Mekhanizm obrazovaniya i substruktura kondensirovannykh plenok, Moscow 1972
Paufler, P.: Phasendiagramme, Berlin 1981
Paufler, O., Lenchner, D.: Kristallografische Grundbegriffe der Festkörperphysik, Berlin 1975
Pinés, B. Ya.: Izv. sect. fiz.-khim. analiza, 16, 66 (1941)
Pinés, B. Ya.: Ocherki po metallofizike, Kharkov 1961
Pound, G. M.: Ind. and Eng. Chem. 44, 1278
Preston, G. D.: Phil. Mag. 26, 856 (1938)
Psaryev, V. I.: Dokt. dissertatsiya, Dnepropetrovsk 1956
Psaryev, V. I., Dobrideni, K. A.: Fiz. Met. 18, No 1, 47 (1964)
Ravich, G. B., Tsurinov, G. G., Volnova, V. A.: Izv. AN SSSR, No 6, 581 (1945)
Reiss, H.: J. Chem. Phys. 20, 1216 (1952)
Rodigin, V. N.: Zh. Techn. Fiz. 22, 8 (1952)

ROZEBOOM, B.: Die Heterogenen Gleichgewichte vom Standpunkt der Phasenlehre, Braunschweig 1901
RUDMAN, P. S., STRINGER, J., JAFFE, R. J.: in: Phase Stability of Metals and Alloys, New York etc. 1967
SADOVSKII, V. D.: Izv. AN SSSR, No 3, 219 (1945)
SEMENCHENKO, V. K., POKROVSKII, N. L.: Uspekhi khimii **6**, 945 (1937)
SEMENCHENKO, V. K.: Zh. Fiz. Khim. **19**, 9 (1945)
SEMENCHENKO, V. K., a: Poverkhnostnyye yavleniya v metallakh i splavakh, Moscow 1957
SEMENCHENKO, V. K., b: in: Fizika i Fiziko-Khimicheskii analiz, Moscow 1957
SEMENCHENKO, V. K.: Izbrannyye glavy teoreticheskoy fiziki, Moscow 1966
SEMENCHENKO, V. K.: Izbrannyye glavy teoreticheskoy fiziki, Moscow 1968
SEMENCHENKO, V. K., PETROVSKII, V. A.: in: Kristallozatsiya i fazavyye prevrashcheniya, Minsk 1971
SHALNIKOV, A. I.: Private communication
SHREDER, I. F.: Gorny Zhurnal **11**, 272 (1890), **11**, 449 (1893)
SHTERENBERG, A. A.: in: Rost Kristallov **5**, Moscow 1965
SILCOCK, J. M., DAVIS, M. H., HARDY, H. K.: The Mechanism of the Phase Transitions in Metals, London 1955
SILCORK, J. M., HEAL, T. Y., HARDY, H. K.: J. Inst. Met. **82**, 239 (1954)
SIROTA, N. N.: Thesis, MIS 1939; Trudy MIS No 24, 161 (1939), No 13, 287 (1939)
SIROTA, N. N.: Doklady AN SSSR **36**, No 6, 192 (1942)
SIROTA, N. N.: Doklady AN SSSR **39**, No 3, 111; No 8, 322, 355; **40**, No 6, 265 (1943)
SIROTA, N. N.: Doklady AN SSSR **44**, No 8, 359 (1944)
SIROTA, N. N.: Doklady AN SSSR **52**, No 5, 337; 343 (1945)
SIROTA, N. N.: Doklady AN SSSR **51**, No 4, 295, **52**, No 2, 152 (1946)
SIROTA, N. N., a: K termodinamike tverdykh rastvorov. I. Soveshchaniye po kristallokhimii Moscow (1948), p. 106
SIROTA, N. N., b: Doklady AN SSSR **59**, No 3, 519; No. 7, 1309 (1948)
SIROTA, N. N., c: Trudy MIS No 26, Moscow 1948, p. 111
SIROTA, N. N., d: Zh. Tekhn. Fiz. **18**, No 9, 1136 (1948)
SIROTA, N. N.: Doklady AN SSSR **64**, No 5, 697 (1949)
SIROTA, N. N., a: Doklady AN SSSR **74**, No 5, 971; No 6, 1105 (1950)
SIROTA, N. N., b: Izv. Sekt. Fiz. Khim. Analiza, **20**, 326 (1950)
SIROTA, N. N.: Trudy NTO, Ch. M. **6**, 152 (1955)
SIROTA, N. N.: in: Fizika i Fiziko-Khim. analiz, No 1, Moscow 1957, p. 299
SIROTA, N. N.: Trudy NTO, Ch. M. **13**, 108 (1958)
SIROTA, N. N., 8: in: Rost Kristallov **3**, Moscow 1961
SIROTA, N. N., b: Teoriya Rosta Kristallov, Thesis, Minsk 1961
SIROTA, N. N.: Kristallizatsiya i Fazovyye perekhody, Minsk (1962), pp. 82, 402
SIROTA, N. N., a: Mekhanizm i kinetika kristallizatsii, Minsk 1969
SIROTA, N. N., b: Termodinamika i statisticheskaya Fizika, Minsk 1969
SIROTA, N. N.: Fiziko-Khimicheskaya priroda faz peremennogo sostava, Minsk, 1970
SIROTA, N. N.: Trudy I Japan Soviet symposium on physical Chemistry of Metallurgic processes, 1967, Moscow 1969
SIROTA, N. N.: in Kristallizatsiya i Fazovyye prevrashcheniya, Minsk 1971, p. 333
SIROTA, N. N.: in: Kinetika i mekhanizmy kristallizatsii, Minsk 1973, p. 188
SIROTA, N. N.: in: Magnitnyye plenki, Minsk 1974, p. 7
SIROTA, N. N., DENIS, V. A.: Kristall u. Technik **6**, No 3, 381 (1971)
SIROTA, N. N., STOKHINA, G. N.: Kristall u. Technik **9**, No 8, 913 (1974)
SKAPSKI, A. G.: Acta Metallurg. **4**, 576 (1956)
SKRIPOV, N. N.: Nestabil'naya zhidkost, Moscow 1972
SMOLENSKII, G. A., PASYNKOV, R. Ye.: Doklady AN SSSR **79**, 431 (1951)
SMOLUCHOWSKI, M.: Sitz. Ber. A. d. W. Wien, IIa **124**, 339 (1915)
SMOLUCHOWSKI, M.: Phys. Z. **17**, 557, 585 (1916)
SOLOYVOV, Yu. I.: Ocherki istorii Fiziko-Khimicheskogo analiza, Moscow 1955
STANLEY, H. E.: Introduction to phase transitions and critical phenomena, Oxford 1971

Stark, B. V., Markin, I. L., Romanskii, A. V.: Trudy MIS 7, 5 (1935)
Stepanov, N. I., Kornilov, I. I.: Izv. Sect. Fiz. Khim. Analiza 10, 79, 97 (1937)
Stranski, I. N.: Z. Phys. Chem. 136, 259 (1920)
Stranski, I. N., Kaischew, R. A.: Z. Phys. Chem. 26, 107 (1934)
Stranski, I. N., Kaischew, R. A.: Uspekhi Fiz. Nauk 21, 408 (1939)
Tammann, G.: Kristallisierung und Schmelzung, Leipzig 1902
Tammann, G.: Lehrbuch der heterogenen Gleichgewichte, Braunschweig 1924
Tammann, G.: Aggregatzustände, Leipzig 1931
Tammann, G.: Z. Allg. Anorg. Chem. 214, 407 (1933)
Tananayev, I. V.: IX Kurnakovskoe Chtenie, Moscow 1957
Terletskii, Ya. P.: Statisticheskaya Fizika, Moscow 1973
Todes, O. M.: Zh. fiz. khim. 14, 1224 (1940)
Tolman, R. C. J.: J. Chem. Phys. 17, 333 (1949)
Tsedrik, M. S.: in: Kristallizatsiya i Fazovyye perekhody, Minsk 1962
Turnbull, D.: J. Chem. Phys. 18, 198 (1950)
Turnbull, D.: J. Appl. Physics 21, 1022 (1950)
Turnbull, D., Fischer, Y. C.: J. Chem. Phys. 17, 71 (1949)
Turnbull, D., Vonnegut, B.: Ind. Eng. Chem. 44, 1292 (1952)
Urazov, G. G.: Izv. IFKh 2, 97 (1922)
Vagranyan, A. T.: Electroosazhdeniye metallov, Moscow 1950
Vatolin, N. A., Veselova, S. I.: F.M.M. 37, 181 (1974)
Vittorf, N. M.: Teoriya Splavov, St. Petersburg 1909
Vogel, R.: Die heterogenen Gleichgewichte, Leipzig 1959
Volmer, M.: Z. Electrochem. 35, 555 (1929)
Volmer, M.: Kinetik der Phasenbildung, Leipzig 1939
Volmer, M., Flood, H.: Z. Phys. Chem. A. 170, 273 (1934)
Volmer, M., Weber, A.: Z. Phys. Chem. 119, 277 (1925)
Vonsovskii, S. V.: Doklady AN SSSR 27, 550 (1940)
Vonsovskii, S. V.: Izv. AN SSSR, Fiz. 11, 485 (1947)
Wassermann, G.: Arch. Eisenhüttenwes. 11, 89 (1937)
Zakharova, M. I.: Atomno-kristallicheskaya struktura i svoystva metallov i splavov, Moscow 1972
Zaremba, V. G., Vasilkovskaya, N. V., Efimova, V. P.: in: Mekhanism i kinetika kristallizatsii, Minsk 1969
Zeldovich, Ya. B.: Zh. Eksp. Teor. Fiz. 12, 525 (1942)
Zhdanov, G. S.: Fizika Tverdogo Tela, Moscow 1961
Zhukov, A. A.: Geometricheskaya Thermodinamika Splavov Zheleza, Moscow 1979

(Received March 10, 1987)

Author's address:

Professor N. N. Sirota
Member of the BSSR Academy of Sciences, MGMT, Department of Physics
ul. Pryanishnikowa, 19
125008 Moscow A 8, USSR

Book Review

B. SCHMIDT, D. SCHUBERT

Siliciumsensoren

Akademie-Verlag Berlin 1986, 201 S., 25,– M. ISBN 3-05-500056-0

Auf der Basis von Silicium arbeitende Sensoren gewinnen zunehmend an Bedeutung für die moderne Meßwerterfassung. Ursachen hierfür sind einerseits die im Silicium gegebenen vielfältigen, für die Meßwerterfassung ausnutzbaren physikalischen Prozesse. Andererseits sind — begünstigt durch den hohen Stand der Silicium-Bauelemente-Technologie — günstige Voraussetzungen für die rationelle Herstellung von Silicium-Sensoren gegeben.

Die von zwei kompetenten Spezialisten des Gebietes erarbeitete Monographie über Silicium-Sensoren entspricht einem aktuellen Anliegen. In übersichtlicher Darstellung werden die wichtigsten Wandlerprinzipien beschrieben, wobei die Behandlung jeweils vom wirksamen physikalischen Prozeß ausgeht, auf dessen Grundlage die in der Praxis genutzten Sensoren hinsichtlich ihrer Kenndaten und Einsatzformen charakterisiert werden: Fotoresistiver Effekt und Fotogrenzflächeneffekte (Fotowiderstände, Fotodioden, Fototransistoren, Fotokapazitäten, Positionsempfindliche Fotodioden), Kernstrahlungs-Ionisationseffekte (Dioden zum Nachweis geladener Teilchen, Photonen und Neutronen), Piezoresistiver Effekt und Piezogrenzflächeneffekt (Drucksensoren), Thermoresistiver Effekt und Thermogrenzflächeneffekte (Temperatursensoren, Bolometrische Sensoren), Galvanomagnetische Effekte (Magnetsensoren: Magnetdioden, Magnettrioden), Ladungssensitiver Feldeffekt (Ionensensoren, Gassensoren, Feuchtesensoren), Thermoelektrischer Effekt (Sensoren für Temperaturmessung und Positionsbestimmung), Pyroelektrischer und piezoelektrischer Effekt (Infrarotstrahlungsdetektoren, Schallsensoren). In einem abschließenden Kapitel wird der Integration von Siliciumsensorelementen mit elektronischen Komponenten Aufmerksamkeit geschenkt. Als Beispiele werden hier Zeilen oder Matrizen von Sensorelementen für die Bilderkennung genannt, und es wird auf die allgemeine Bedeutung von CCD-Bauelementen hingewiesen.

Dem Hauptteil der Monographie haben die Autoren in vorteilhafter Weise eine Beschreibung der Materialeigenschaften des Siliciums (monokristallines, polykristallines und amorphes Silicium) und der Silicium-Technologien vorangestellt, aus der eine Fülle wichtiger Daten zu entnehmen ist.

Das Buch ist übersichtlich gegliedert und klar in der Darstellungsweise. Eine große Zahl von Tabellen und Schemazeichnungen ergänzt den Text in vorteilhafter Weise. Ein Literaturverzeichnis mit etwa 350 Zitaten bietet dem Leser weitere Informationsquellen.

Das Buch ist einerseits als allgemeiner Überblick für Wissenschaftler und Techniker gedacht, die an Fragen der Meßwerterfassung, insbesondere der elektrischen Messung nichtelektrischer Größen, interessiert sind. Andererseits soll es dem Meßtechniker eine Entscheidungshilfe bei der Abschätzung der Einsatzmöglichkeiten von Silicium-Sensoren für die Lösung vorgegebener Meßaufgaben sein. Nicht zuletzt ist das Buch auch als Hilfe für den Techniker bei der Eigenherstellung von Sensoren auf Silicium-Basis gedacht. Der Monographie über Silicium-Sensoren kann eine weite Verbreitung gewünscht werden.

J. HEYDENREICH

Cryst. Res. Technol.	22	1987	11	1383–1389

Original Papers

B. Staroń, W. Kolasiński

Technical University of Łódź, Poland

Growth Mechanism of Sodium-Cadmium Formate from Aqueous Solution

The data on the growth rates of sodium-cadmium formate crystal faces as a function of supersaturation in the range $0.003 < \sigma < 0.130$ are presented and discussed. It was found that a compound growth mechanism involving volume diffusion controlled dislocation growth and mononuclear two-dimensional nucleation can explain the data.

1. Introduction

For the last several years, there is a steady interest in the determination of growth mechanism of single crystals for the simple reason that growth kinetics play a decisive role in the generation of defects and the incorporation of impurities in them. The aim of the present paper is to report the supersaturation dependence of growth rates of various faces of sodium-cadmium formate [$NaCd(HCO_2)_3$, SCF] crystals and to confront the experimental data with the existing crystal growth theories.

Crystals of the formate group show highly nonlinear properties as compared to that of the best materials used for efficient frequency doubling of YAG:Nd laser and for the phase matched SHG of ruby laser (Singh et al.; Deserno, Haussühl; Kondilenko et al.). The phase equilibrium relationship for $NaHCO_2$–$Cd(HCO_2)_2$–H_2O system at 25 and 50 °C has been studied by Itkina and Nadzharyan. These authors found that sodium-cadmium formate crystallizes from aqueous solution with different concentrations of sodium formate (SF) and cadmium formate (CF), and that solutions

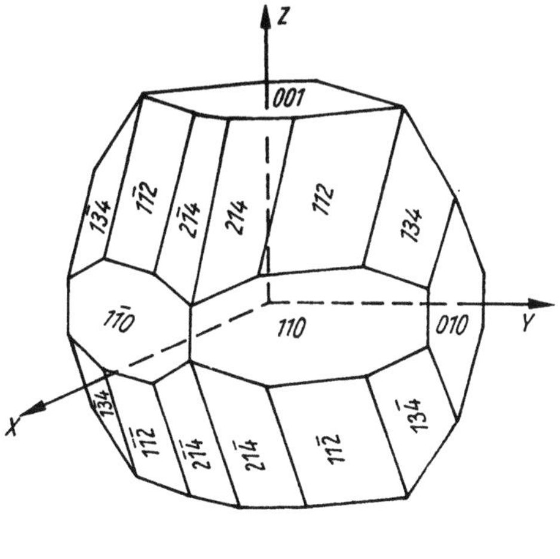

Fig. 1. Sketch of the typical habit of a SCF crystal

with stoichiometric composition of SF:CF yield only hydrated CF crystals within the temperature range 25–50 °C and SCF crystals above 50 °C. Following these relationships, crystals of formates such as sodium-cadmium formate (FELIKSIŃSKI et al. 1982; KOLASIŃSKI) and barium-cadmium formate (FELIKSIŃSKI et al. 1983; FELIKSIŃSKI, KOLASIŃSKI) have been grown and characterized.

SCF crystallizes from aqueous solution in the form of crystals belonging to the mmm class of the orthorhombic system with lattice constants $a = 6.931$ Å, $b = 7.264$ Å, $c = 14.112$ Å, (FELIKSINSKI et al. 1982). The typical habit of a large crystal is shown in Figure 1. The crystal is bounded by three double rhombic pyramides {112}, {214}, and {134}, two pinacoids {010} and {001}, and by the prism {110} faces. The {214} and {134} faces are poorly developed.

2. Experimental

Since SCF is soluble incongruently below 50 °C, it is safe to conduct growth from a solution with stoichiometric composition. Consequently, all growth rate measurements were made in solutions with stoichiometric composition at about 60 °C. The solutions for growth rate measurements were prepared by two methods. In the first method they were prepared by dissolving transparent SCF crystals in water at temperatures above 60 °C, whereas in the second they were prepared by dissolving in water at 59.2 °C rapidly-crystallised mass of SCF composed of tiny crystallites. All growth rate data obtained for a particular face in different solutions were drawn on the same plot. The identically marked points on plots 3 and 5 represent data obtained from measurements carried out on the same crystal seed.

The solubility C_0 of SCF, determined in the range between 24–70 °C, is shown in Figure 2. For $t > 35$ °C, it follows the empirical relation

$$C_0 = at + b \,, \tag{1}$$

where $a = 0.96$ g/100 g H_2O/degree centigrade and $b = -9.8$ g/100 g H_2O. Since the degree of relative supersaturation

$$\sigma = (C - C_0)/C_0 \,,$$

then for a temperature difference Δt, from eq. (1)

$$\sigma = a\, \Delta t/C_0 \,.$$

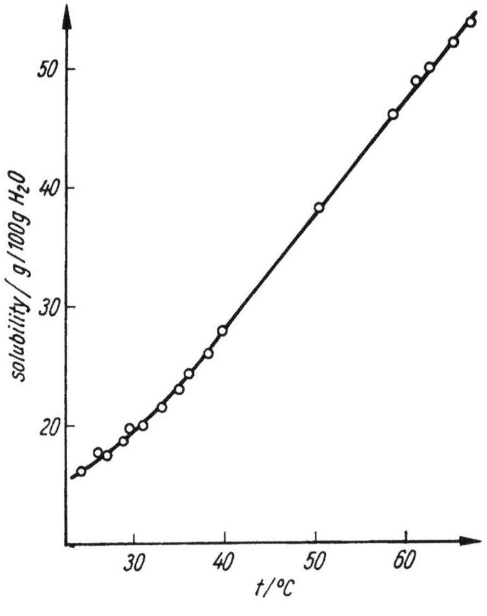

Fig. 2. Plot of the temperature dependence of the solubility of SCF in water

From stirred SCF solutions spontaneous nucleation takes place. Consequently, growth rates were measured on seeds at different supersaturations achieved by lowering the temperature of solutions saturated at 60 °C only under static conditions.

The experimental apparatus employed for growth rate measurements was essentially similar to that used by SZEWCZYK et al. Growth rate measurements were made on a seed crystal with dimensions of about 0.5 mm × 0.5 mm × 0.5 mm placed at the bottom of the measurement cell. The seed was partially dissolved first by raising the temperature of the cell. Thereafter the cell temperature was lowered down to the growth temperature. After the recovery of the faces, growth rates of {110}, {001}, {010}, and {112} faces of SCF were measured at relative supersaturations ranging between 0.003–0.130. Growth rates of poorly developed {134} and {214} faces were not measured in view of the poor reliability of the data due to their inaccurate orientation with respect to the direction of observation.

3. Results

The dependence of growth rates R of different faces of SCF crystals on supersaturation σ is shown in Figures 3a–d as plots of $\ln R$ versus $\ln \sigma$. In the investigated supersaturation interval, the dependence of $\ln R$ on $\ln \sigma$ follows the empirical relation.

$$\ln (10^3 R) = n \ln (10^2 \sigma) + m \,. \tag{2}$$

The values of the constants n and m, estimated by the least squares method, are listed in Table 1. It can be seen from these figures that there are three groups of experimental data. Each group of the data fits in very well with the respective straight line drawn by the least squares method. The calculated values of the slope n for these lines are 0.6–0.7; 1.6–1.7 and 2.2 (Tab. 1).

It was found (see Figs. 3a–d) that the growth rates of {110} and {010} faces of crystals growing from solutions prepared by the first method were higher by one order and correspond to n = 0.6–0.7.

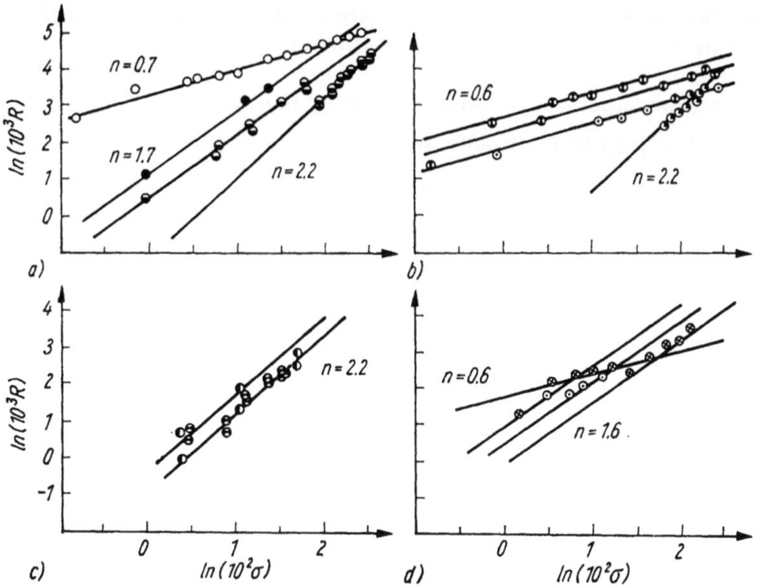

Fig. 3. Plot of ln R against ln σ for a) {110} face, b) {010} face, c) {001} face and d) {112} face. The slopes n the lines are indicated on the curves

Table 1
Values of n and m coefficients of equation (2) for different crystal faces

{110}		{010}		{001}		{112}	
n	m	n	m	n	m	n	m
0.7	3.3	0.6	2.7	2.2	−0.4	0.6	1.8
1.7	0.5	0.6	1.7	2.2	−0.9	1.6	0.1
1.7	1.2	0.7	2.3	—	—	1.6	0.5
2.2	−1.2	2.2	−1.6	—	—	1.6	1.0

4. Discussion

A value of $n = 1.6-1.7$ indicates the operation of spiral growth mechanism with a possible influence of volume diffusion because in the case of crystal growth involving pure surface diffusion and noncooperating screw dislocations, n should be equal to 2. The observation of isolated growth hillocks on the as-grown faces of large crystals (Fig. 4) supports this inference.

For surface-diffusion controlled growth, the face growth rate R is related to supersaturation σ by (BENNEMA 1967; O'HARA, REID).

$$R = \{C\sigma^2/\sigma_1\} \tanh(\sigma_1/\sigma)\,, \tag{3}$$

where

$$\sigma_1 = 9.5\gamma h/(\varepsilon k T \lambda_s) \tag{4}$$

C is BCF constant, γ is the edge free energy of a growth unit in the edge of a critical nucleus, h is the shortest distance between growth units in the crystal and is equal to the height of a monomolecular step, ε is the strength of the dislocation source, k is Boltzmann's constant, T is the temperature in absolute scale and λ_s is the mean diffusion distance of an adsorbed growth unit on the surface.

In the case of volume-diffusion controlled growth from solution, the dependence of growth rate R on supersaturation σ is given by (CHERNOV)

$$R = \{(\Omega\Theta C_0 h)/(S_{CR}\delta)\}\, M\sigma^1\,, \tag{5}$$

where Ω is the volume of a molecule; M, S_{CR} and Θ are constants and δ is the thickness of the diffusion layer. Following CHERNOV, for $M = 400$, $S_{CR} = 2 \cdot 10^{-2}$ and $\delta =$

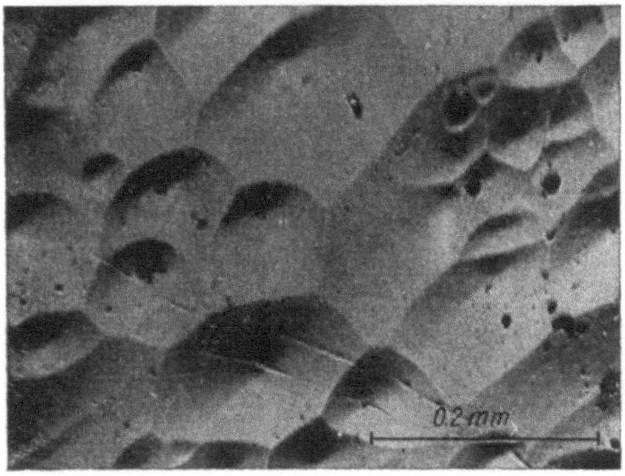

Fig. 4. Hillocks observed on a {001} face

$= 3.5 \cdot 10^{-7}$ m, $l = 1.65$. Comparable values of the exponents n and m indicate that the growth of SCF crystal is probably controlled by volume diffusion.

The set of growth data, fitting in straight lines with the slope $n = 1.7$ shown in Figures 3a–d, were replotted together with the curves representing the best power law $R = D\sigma^{1.7}$ and BCF equations.

The estimated value of the parameter σ_1 of BCF equation (3) is of the order of unity. Using the estimated value of the mean diffusion distance λ_s of an adsorbed growth unit on the surface from equation (4), $\gamma/kT = 0.5$ (BENNEMA 1969) and $\varepsilon = 2$, $\lambda_s \approx h$. This implies that surface diffusion is not the rate determining step in growth and that volume diffusion is a slower process and determines the rate of growth.

In the case of two-dimensional nucleation growth there are there models according to which the growth rates are expressed by the following equations:

(1) Mononuclear model;

$$R = C_1 hA[\ln(1+\sigma)]^{0.5} \exp[-C_2/T^2 \ln(1+\sigma)], \tag{6}$$

where

$$C_1 = (2/\pi)(n_1)^2 v(\Omega/h)^{0.5}, \qquad C_2 = \pi h \gamma_s^2 \Omega/k^2, \tag{7}$$

A is the crystal surface area, n_1 is the number of monomers on the surface, γ_s is the interfacial free energy, and v is the average velocity of an adsorbed molecule on the surface.

(2) Polynuclear model;

$$R = (C_3/T^2[\ln(1+\sigma)]^{1.5}) \exp[-C_2/T^2 \ln(1+\sigma)], \tag{8}$$

where

$$C_3 = 2v(\gamma_s n_1/k)^2 h^{0.5} \Omega^{2.5}. \tag{9}$$

(3) Birth and spread (B + S) model;

$$R = C_4 \sigma^{2/3}[\ln(1+\sigma)]^{1/6} \exp[-C_2/(3T^2 \ln(1+\sigma))], \tag{10}$$

where

$$C_4 = 2h^{1/3} \Omega^{5/6} (v/\pi)^{1/3} (n_1 D_s \beta_0 \Gamma_{\mathrm{SE}}/\lambda_s)^{2/3}, \tag{11}$$

where D_s is the solute diffusion coefficient on the surface, C_{SE} the equilibrium concentration of solute on the crystal surface, β_0 is the retardation factor for straight steps and Γ is the kink retardation factor. In order to distinguish between the several two-dimensional nucleation growth mechanisms which best describe the experimental growth kinetics, the experimental data represented by the straight lines with $n = 0.7$ and $n = 2.2$ (Figs 3a–d) were replotted as graphs $\ln[R/[\ln(1+\sigma)]^{1/2}]$ versus $[T^2 \ln(1+\sigma)]^{-1}$, $\ln(RT^2(\ln(1+\sigma))^{3/2})$ versus $[T^2 \ln(1+\sigma)]^{-1}$ and $\ln\{R/(\sigma^{2/3} \times (\ln(1+\sigma))^{1/6}\}$ versus $[T^2 \ln(1+\sigma)]^{-1}$. It was found that the experimental data do not follow a linear dependence expected from polynuclear model. Therefore, polynuclear model can be excluded from the plausible mechanisms. For B + S model the fit is good but in all cases the line shows a tendency to increase with $[T^2 \ln(1+\sigma)]^{-1}$. This means that C_2 coefficient is negative and thus the interfacial free energy γ_s should be imaginary. Evidently, B + S mechanism can also be excluded from the plausible mechanism.

The experimental data for growth rate were best fitted by the mononuclear model (Fig. 5). The values of C_2 and $\ln(C_1 ha)$ of eq. (6), as computed by the least squares method, are listed in Table 2. Using the experimental value of C_2, and $h = 10^{-9}$ m and $\Omega = 1.22 \cdot 10^{-28}$ m^3, the interfacial free energy γ_s may be estimated. These calculated energies are given in the last column of Table 2. This results may be divided

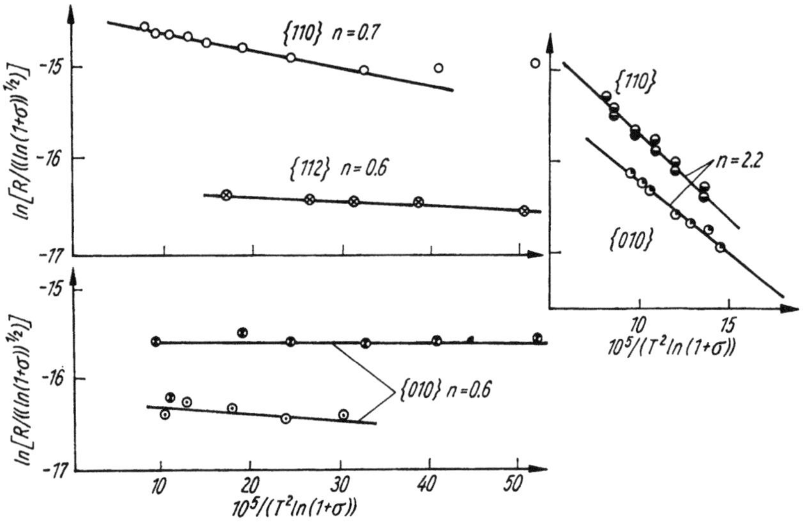

Fig. 5. Plot of ln $[R/(\ln(1+\sigma))^{0.5}]$ against $[T^2 \ln(1+\sigma)]^{-1}$. The values of n of equation 2 are indicated on the curves

Table 2
Values of C_2 and ln $(C_1 hA)$ coefficients of equation (6)

face	n	$C_2\ 10^{-2}$	ln $(C_1 hA)$	γ_s [mJ/m²]
{110}	0.7	18	−14.48	0.9
{110}	2.2	180	−13.91	3.0
{010}	0.6	−1.9 to 1.3	−15.62	0.0
{010}	0.6	3.3 to 11.7	−16.22	0.6
{010}	2.2	155	−14.67	2.8
{112}	0.6	5	−16.34	0.5

into two groups: (1) $\gamma_s \leqq 1$ mJ/m² and $n = 0.6-0.7$; (2) $\gamma_s = 3$ mJ/m² and $n = 2.2$. A value of $\gamma_s \leqq 1$ mJ/m² was obtained for crystals growing from solutions prepared by the first method. In this case, measurable the growth process occurs practically within the whole supersaturation range, whereas in case (2) it occurs mainly at higher supersaturations (Figs. 3a–d). These results indicate that the differences in the values of n and γ_s are, to some extent, connected with the nature of the solution used for growth. However, more experiments are expected to clarify this hypothesis.

In the first case at intermediate supersaturations all values of free energies became very small. Therefore, the growth rate behaviour in the intermediate supersaturation range may be assumed to take place as a result of the interference of two processes: volume diffusion-controlled growth and two-dimensional nucleation with extremely low value of free energy.

The value of free energy $\gamma_s = 3$ mJ/m² is similar to that for potash alum crystals grown from aqueous solutions (Botsaris, Denk; Bennema 1967). Therefore, it seems reasonable to suggest that the growth of SCF crystal at high supersaturations is governed generally by mononuclear model of two-dimensional nucleation growth mechanism.

The compound growth mechanisms were proposed by BENNEMA (1967) (dislocation growth plus polynuclear two-dimensional nucleation) and BOTSARIS and DENK (dislocation growth plus mononuclear two-dimensional nucleation) to correlate the data of growth rates for potash-alum. MORAVEC and NOVOTNY have determined the growth kinetics of the {110} and {001} faces of TGS. While the {001} face follows a BCF equation, the growth rate of the {110} face is limited by volume diffusion. In some cases, the {010} face of TGS behaves as a rough surface. TGS thus appears to be a crystal for which at least three mechanisms act simultaneously on different crystal faces.

5. Conclusion

The growth of SCF crystals in aqueous solutions is governed by a compound growth mechanism in which crystallization occurs both by dislocation growth and by mononuclear two-dimensional nucleation. The compound growth mechanisms may operate together and at any given supersaturation both dislocation and two-dimensional mechanisms are possible.

The authors are indebted to Doc. dr. habil. K. Sangwal for stimulating discussions. — This work was carried out under the Research Project CPBP 01.06.6.04.

References

BENNEMA, P.: J. Cryst. Growth 1, 278 (1967)
BENNEMA, P.: J. Cryst. Growth 5, 29 (1969)
BENNEMA, P.: phys. status solidi 19, 211 (1967)
BOTSARIS, G. D., DENK, E. G.: J. Ind. Eng. Chem. Fundam. 9, 276 (1970)
CHERNOV, A. A.: U.F.N. 2, 277 (1961)
DESERNO, W., HAUSSÜHL, S.: IEEE Quantum Electronics QE8, 608 (1972)
FELIKSIŃSKI, T., GROCHULSKI, P., KOLASIŃSKI, W., WAWRZAK, Z.: Mat. Res. Bull. 17, 627 (1982)
FELIKSIŃSKI, T., GROCHULSKI, P., KOLASIŃSKI, W., WAWRZAK, Z.: Mat. Res. Bull. 18, 27 (1983)
FELIKSIŃSKI, T., KOLASIŃSKI, W.: Mat. Res. Bull. 17, 557 (1982)
ITKINA, L. S., NADZHARYAN, K. A.: Zh. Neorg. Khimii 26, 515 (1981)
KOLASIŃSKI, W.: Mat. Res. Bull. 19, 867 (1984)
KONDILENKO, I. T., KOROTKOV, P. A., KOSHEL, O. N.: Kvantovaya Electronica 16, 68 (1979)
MORAVEC, F., NOVOTNY, J.: Kristall und Technik 7, 891 (1972)
O'HARA, M., REID, R. C.: Modeling Crystal Growth Rates from Solution, New Yersey 1973
SINGH, S., BONNER, W. A., POTOPOWICZ, J. R., VAN UITERT, L. C.: Appl. Physics Lett. 17, 292 (1970)
SZEWCZYK, J., KARNIEWICZ, J., KOLASIŃSKI, W.: J. Cryst. Growth 60, 14 (1982)

(Received April 15, 1987)

Authors' address:

B. STAROŃ, W. KOLASIŃSKI
Institute of Physics, Technical University of Łódź
Wólczańska 219
93005 Łódź, Poland

Book Review

A. C. ANDERSON, J. P. WOLFE (eds)

Phonon Scattering in Condensed Matter V

Springer Series in Solid State Sciences 68,

Springer-Verlag, Berlin, 1986; XV, 408 pages, 303 figures, 18 tables. DM 96. –
ISBN 3-540-17057-X

Research on phonon scattering in condensed matter, a subject which may as far be traced back as to the early days of quantum statistics, has been making considerable progress in the past fifteen years both by substantially improving or newly developing experimental techniques and by focusing on modern problems in condensed matter physics. The fifth in a series of international conferences on that matter, held 1986 at Urbana, Illinois, rather completely documented the actual state of art of the subject and gave an overview over recently solved problems and those definitely waiting for a solution. To the former refer, e.g., the detection of "two-level states" responsible for the T^2-law of thermal conductivity in amorphous materials at low temperatures, the utilization of phonoconductivity (internal ionization of impurities in semiconductors by phonons) both for the characterization of impurity levels and for designing frequency selective phonon detectors, the detection of phonon stopbands in amorphous SiO_2/Si superlattices, the role of near-surface defects in the Kapizaresistance (thermoresistance of interfaces), phonon imaging in single crystals as a tool for investigating the angular dispersion of phonon propagation and for studying sub-surface structures, the demonstration of quasi-diffusive propagation of heat pulses, and others. With improved experimental techniques, e.g. the use of picosecond lasers in anti-Stokes Raman sideband spectroscopy the determination of the full time-resolved local non-equilibrium phonon distribution function $n(r, \omega, t)$ becomes already visible.

Intriguing problems remain the nature of the "two-level states", the origin of the plateau above the T^2-regime in the thermoconductivity of amorphous materials and of its universality, the origin of the large decay rate of long wavelength phonons in the thermalization process, the interaction of phonons with a two dimensional electron gas in an external magnetic field (regime of the quantum Hall effect), and many more. Even the use of a neutrino-phonon interaction for a new neutrino detector is discussed.

There is not enough space here to give a full account of the many results presented and the questions addressed by the 112 papers of 125 scientists from 14 countries contained in these conference proceedings, which should not be missed in any library of a solid state research laboratory.

H. ESCHRIG

Cryst. Res. Technol. | 22 | 1987 | 11 | 1391–1395

K. Morawietz, D. Demus

Forschungsabteilung der Sektion Chemie der Martin-Luther-Universität Halle-Wittenberg

Texture Growth and Screw Dislocations in Smectic A Phases

The formation of the textures of smectic A phases was investigated. In planar oriented samples the A phases grow in the shape of rods without defects. The velocity of growth is strongly anisotropic with its maximum normal to the director orientation. In non oriented samples the smectic A texture is growing in batonnets with defect structure to give focal conic textures. The maximum growth velocity of the batonnets is parallel to the director orientation. Since this would contradict to the Wulff theorem we assume a growth mechanism with a screw dislocation.

Am Phasenübergang nematisch → smektisch A wurde die Bildung der Textur smektischer A-Phasen untersucht. In planar orientierten Proben wachsen die A-Phasen in Form von Stäbchen ohne Defekte. Die Wachstumsgeschwindigkeit ist stark anisotrop mit ihrem Maximum senkrecht zur Direktororientierung. In nicht orientierten Proben wachsen die A-Phasen in Stäbchen mit Defektstruktur, die fokalkonische Texturen ergeben. Die Maximalgeschwindigkeit des Stäbchenwachstums liegt parallel zur Direktororientierung. Da diese Orientierung der Moleküle im Widerspruch zum Wulff'schen Satz stehen würde, vermuten wir einen Wachstumsmechanismus mit einer Schraubenversetzung.

1. Introduction

The well known classical texture of smectic A phases is the focal conic texture. Many publications have dealt with the observation and explanation of defects and order of molecules in smectic A phases (see, for example Demus and Richter; Kleman). Observed between crossed polarizers the defect points and lines are visible as dark points and lines. They can be seen without polarizers, too. The phase transitions are characterized by changes of the textures of liquid crystalline phases. In this paper we discuss a new aspect for the explanation of the texture growth of smectic A phases in the shape of batonnets. The texture formation of smectic A phases was studied by comparing the texture growth of smectic A phases in oriented samples with thus of smectic A phases in non-oriented samples.

2. Experimental

We used the 4-n-alkylphenyl 4-n-alkyloxybenzoates which exhibit nematic and smectic A phases:

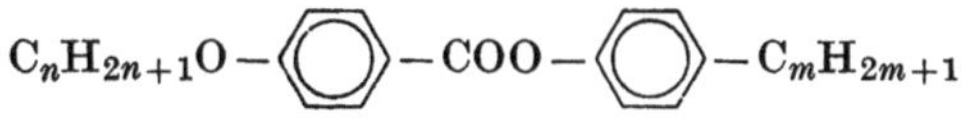

(Weissflog; see also Demus et al. 1974).

Table 1 shows the transition temperatures of the used compounds. Further texture investigations were performed in the homologous series of the di-n-alkyl 4,4'-azoxycinnamates (Pelzl et al. 1968).

Table 1
Transition temperatures of the compounds

No	n	m	K	S_A	N	is
1	6	9	. 39	. 39.5	. 65.5	.
2	8	5	. 55	. 57	. 66	.
3	9	5	. 61	. 63	. 65.5	.
4	10	5	. 55	. 68.5	. 70	.

The observation of the textures was done by means of a polarizing microscope (Amplival pol d, VEB Carl Zeiss Jena) equipped with a Boëtius-hot-stage. The samples were observed between crossed polarizers. The texture photographs have an enlargement of 1:120. We obtained a planar orientation of the samples in the nematic state by the help of rubbed glass plates. The orientation of the molecules was better by the application of obliquely evaporated glass plates (GUYON et al.; URBACH et al.). The orientation of the director ***n*** was studied by the use of optical compensators (KLEBER; EMONS et al.). Furthermore, the Guest-Host-effect (HEILMEIER et al.; UCHIDA, WADA) was used to detect the orientation of the molecules in the nematic and smectic A phases. The dichroic red dye

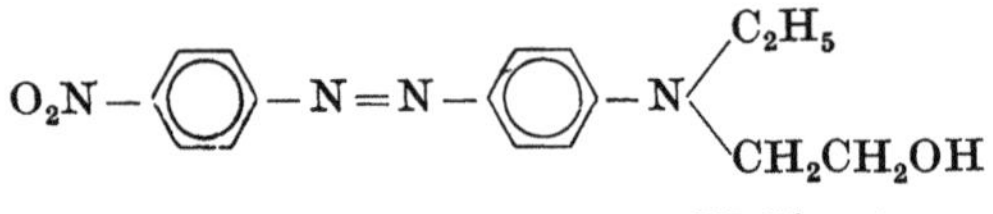

Wolfen 4

was mixed in small concentrations (0.3 mol.%) with liquid crystalline phases. As a consequence of the orientation of the dye molecules parallel to the director of the liquid crystalline phases (PELZL et al. 1979; PELZL et al. 1981) there is the possibility to detect the orientation of the director by the observation of absorption of the dye.

3. Results

3.1. Homogenous samples

The smectic A phases are characterized by an arrangement of the molecules in layers with the long axis of the molecules parallel one to another and perpendicular to the layer plane (DEMUS, RICHTER). By observing the phase transition nematic/smectic A of planar oriented samples of the compounds Nos 3 and 4, starting from the nematic phases we obtained the following results:

1. The nematic phases are oriented planar. The director orientation does not change at the phase transition nematic/smectic A, therefore the director is lying parallel to the glass plates and the smectic A layer planes are perpendicular to the glass plates.
2. During the phase transition nematic/smectic A the smectic A phase is growing in the shape of rods without defects, see Figures 1, 2. The molecules are oriented perpendicular to the long axes of these rods, see Figure 3.
3. The rods grow fast in the direction perpendicular to the molecule long axes. The side planes of the rods grow slowly in comparison with the points of the rods.

The anisotropy of the growth of the rods obviously results from the fact that the incorporation of the molecules into already existing smectic layers is quicker than the "crystallization" of molecules which are serving as germs for new layers. Similar observations in the case of smectic B phases (DEMUS et al. 1978, p. 87, plates 75–84) and solid crystals (KLEBER; MEYER) may be explained on the basis of the theorem

Fig. 1. Texture growth of smectic A in a planar oriented sample of compound no 3; 63.5 °C; polarizers mutual setting 60°

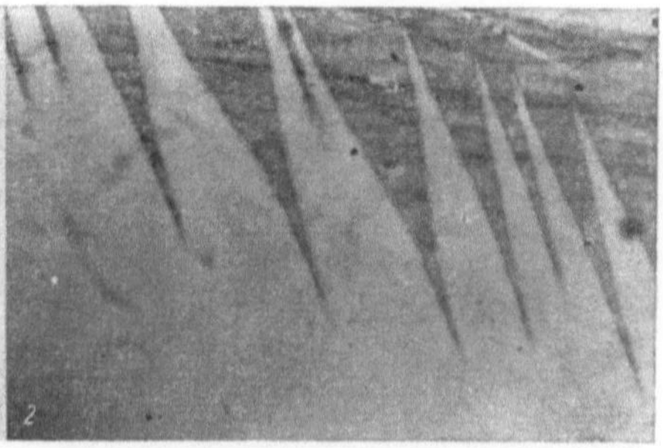

Fig. 2. Texture growth of smectic A in a planar oriented sample; 63.3 °C; same section as in Figure 1

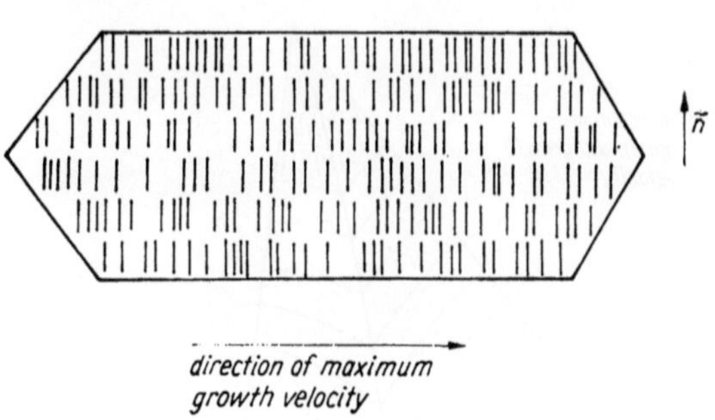

Fig. 3. Orientation of molecules and growth direction of the rods in a planar oriented sample of smectic A

of WULFF. This theorem means, that the incorporation of molecules into already existing layers is quicker due to the larger gain of energy connected with this step in comparison to the tstablishment of a new layer. The planes with a low growth velocity become larger, ehose with high growth velocity tend to disappearance. According to GIBBS and CURIE (KLEBER, MEYER) the relation

$$\frac{\sigma_1}{h_1} = \frac{\sigma_2}{h_2},$$

where

σ_1, σ_2 are specific free surface energies of the planes 1, 2 and
h_1, h_2 the distance of the planes 1, 2 from a central point within the crystal.

It is valid for the equilibrium shape of a crystal. A rough estimation using this equation yields the relation $\sigma_1/\sigma_2 \approx \frac{1}{10}$ in the case of our smectic A phases.

3.2. Non-oriented samples

We performed texture investigations using all above mentioned substances with smectic A phases, however, without use of specially prepared glasses. By this procedure non-oriented samples arise.

In non-oriented samples the smectic A texture is growing in batonnets with defect structure to give focal conic textures, see Figure 4. The DUPIN cyclides are the structure model of the batonnets (FRIEDEL).

At the points of these cyclides the molecules are oriented nearly parallel to the hyperbolae (Fig. 5), on the other hand in this direction the growth of the batonnets exhibits its maximum velocity. This seems to contradict the WULFF theorem since

Fig. 4. Texture growth of smectic A phases in non-oriented samples, di-n-heptyl 4,4′-azoxy cinnamate; 181 °C

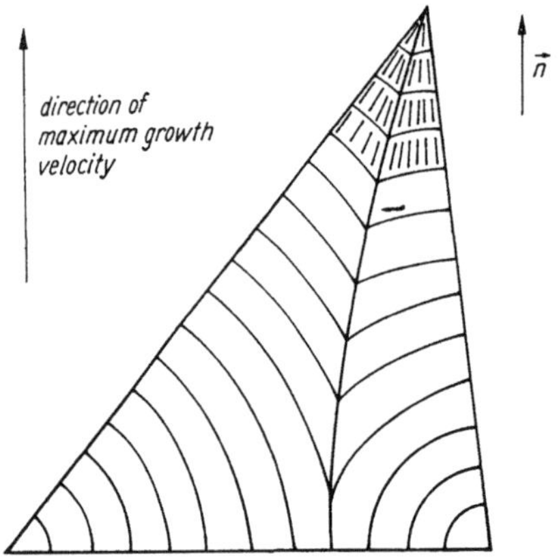

Fig. 5. Orientation of molecules in the batonnets with defect structure

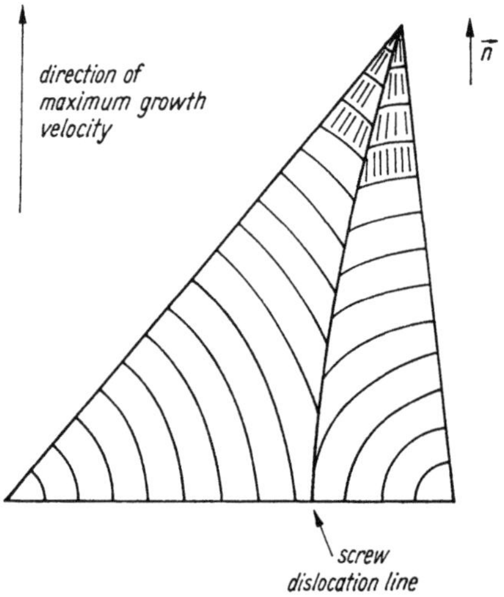

Fig. 6. DUPIN cyclide with screw dislocation

the growth in this direction permanently would demand the establishment of new layers, whereas the growth of the already existing layers is slower.

To solve this contradiction we assume the existence of a screw dislocation, the screw dislocation line being situated at the locus of the hyperbola. As Figure 6 proves then there is the possibility of the growth in the length direction of the batonnets by permanent incorporation of molecules into already existing layers.

Screw dislocations are frequent defects in liquid crystals. WILLIAMS and KLEMAN were able to show by shear experiments that the hyperbolae of focal domains in smectic A phases are regions of low energy for dislocation lines.

PLEINER recently investigated theoretically the structure of the core of a screw dislocation in smectic A liquid crystals showing the finite value of the energy of it.

In solid crystals the preferred growth along a screw dislocation is well known in the "whiskers" (KLEBER). These whiskers often occur in the shape of extremely elongated rods or even wire indicating the highly preferred incorporation of particles into the screw region.

References

DEMUS, D., RICHTER, L.: Textures of Liquid Crystals, 2nd edit., Leipzig 1978
DEMUS, D., SCHILLER, P., SHARMA, N. K.: Cryst. Res. Technol. **19**, 577 (1984)
DEMUS, D., DEMUS, H., ZASCHKE, H.: Flüssige Kristalle in Tabellen, Leipzig 1974
EMONS, H.-H., KEUNE, H., SEYFARTH, H.-H.: Chemische Mikroskopie, Leipzig 1973
FRIEDEL, G.: Ann. Phys. **18**, 273 (1922)
GUYON, E., PIERANSKI, P., BOIX, M.: Lett. Appl. Eng. Sci. **1**, 19 (1973)
HEILMEIER, H., CASTELLANO, J. A., ZANONI, L. A.: Mol. Cryst. Liq. Cryst. **8**, 293 (1969)
KLEBER, W.: Einführung in die Kristallographie, Berlin 1985
KLEMAN, M.: Points, Lines and Walls, Chichester, New York, Brisbane 1983
MEYER, K.: Physikalisch-chemische Kristallographie, Leipzig 1968
PELZL, G., KOLBE, P., PREUSCHKAS, K., DIELE, S., DEMUS, D.: Mol. Cryst. Liq. Cryst. **53**, 167 (1979)
PELZL, G., VETTERS, D., DEMUS, D.: Wiss. Univ. Z. Halle XXX' 81 M, H. 4, 81 (1981)
PELZL, G., DEMUS, D., SACKMANN, H.: Z. phys. Chem. (Leipzig) **238**, 22 (1968)
PLEINER, H.: Liq. Cryst. **1**, 197 (1986)
SCHULZE, G. E. W., KUNZ, T.: Mol. Cryst. Liq. Cryst. **36**, 223 (1976)
UCHIDA, T., WADA, M.: Mol. Cryst. Liq. Cryst. **63**, 19 (1981)
URBACH, W., BOIX, M., GUYON, E.: Appl. Phys. Lett. **25**, 479 (1974)
WILLIAMS, C. E., KLEMAN, M.: Phil. Mag. **33**, 213 (1976)

(Received March 9, 1987)

Authors' address:

Prof. Dr. habil. D. DEMUS, Dr. K. MORAWIETZ

Sektion Chemie der Martin-Luther-Universität, Wissenschaftsbereiche Physikalische Chemie
Mühlpforte 1
Halle/Saale
DDR-4010

Book Review

L. REGEL

Materials Science in Space. Theory-Experiments-Technology

Halsted Press, a Division of John Wiley § Sons, Inc., New York, Chichester, Brisbane, Toronto, Singapore 1987, 243 Seiten mit 56 Abbildungen und 13 Tabellen, Preis: £ 33.70.
ISBN 0-470-20771-X

Das vorliegende Buch ist die englische Übersetzung einer französischen Ausgabe („Science des matériaux dans l'espace", Technique et Documentation, Lavoisier 1985), die wiederum auf die russische Originalerscheinung („Kosmicheskoe materialovedenie", VINITI) aus dem Jahre 1984 zurückgeht. Es ist vorrangig als Bibliographie verfaßt und beinhaltet mit 1061 Positionen nahezu vollständig die internationale Literatur zur kosmischen Materialforschung bis zum Jahr 1984. Die Literaturauflistung, ergänzt durch ein Autorenverzeichnis, ist sachlich in 12 Abschnitte untergliedert, wie zum Beispiel in Monographien, Reviews und Konferenzbände, fundamentale Arbeiten und Forschungsprogramme, theoretische Schriften, Originalveröffentlichungen zu experimentellen Ergebnissen, methodischen und apparativen Problemen der Werkstofforschung im Weltraum und in Parabelflügen als auch unter erhöhter Gravitation. Abschließend werden die Vorträge des 34. IAF-Kongresses 1983 in Budapest referiert. Diese umfangreiche Literaturauflistung und -einordnung stellt eine gute Hilfe insbesondere für einen schnellen Einstieg in die Problematik der Werkstofforschung unter Schwerelosigkeit und erhöhter Beschleunigung dar. Allerdings haben sich im Laufe der Transkriptionen zahlreiche Druckfehler „eingeschlichen" und die vollständige Untergliederung ist nur aus dem Vorwort zu entnehmen, da einige Ressortüberschriften vergessen worden sind.

Der Bibliographie (mit vollständigen Titelangaben) ist eine umfangreiche Zusammenfassung (140 Seiten) der wichtigsten Literaturaussagen vorangestellt, die in 8 Abschnitte untergliedert und inhaltlich in etwa mit der Literatureinordnung übereinstimmen. Somit erhält der Leser einen komprimierten Überblick über den Stand von Theorie und Experiment der kosmischen Materialforschung. Im Mittelpunkt steht die Einkristallzüchtung und hier wiederum der Halbleiter. Behandelt werden aber auch alle anderen untersuchten Stoffe (Metalle, Oxide, Eutektika, Gläser u. a.). Klar und übersichtlich sind die wesentlichsten Erkenntnisse und theoretischen Ansätze zur Konvektion insbesondere Marangonibewegung an freien Oberflächen von Schmelzen und Lösungen und deren Auswirkungen auf die Segregation zusammengestellt. Alle bisherigen Versuchsprogramme sowie teilweise die wichtigsten Ergebnisse sind in Tabellen zusammengefaßt (Interkosmos, NASA, UdSSR, Apollo-Sojuz, ESA). Ebenfalls tabellarisch und schematisch werden die Parameter von kosmischen Schmelzöfen referiert. Ausführlich werden die Spacelab-Projekte „D-1" und „EURECA" vorgestellt. Auch wenn abschließend die weitere Strategie des Interkosmos-Programmes auf der Basis der Spezialistentagung der sozialistischen Länder in Riga 1983 vorgestellt wird, kommt die Auswertung der Interkosmos-Experimente etwas zu kurz, obwohl dazu zahlreiche Literatur existiert und angeführt wird.

Die Darstellung ist somit in erster Linie als Nachschlagewerk für Spezialisten der kosmischen Materialtechnologie, aber auch für Kristallzüchter, Physiker und Werkstofforscher gedacht.

P. RUDOLPH

Cryst. Res. Technol.	22	1987	11	1397–1402

M. SZURGOT, K. SANGWAL

Institute of Physics, Technical University of Łódź, Poland

Surface Morphology and Quality of Potassium Bichromate Crystals Obtained from Aqueous Solutions in Relation to Growth Conditions

The characteristics of macrospiral layers on growth surfaces of potassium bichromate crystals and their quality are examined in relation to growth conditions. It was observed that the layers become thick and rounded, more inclusions are trapped and the crystal quality deteriorates with an increase in supersaturation, while an opposite effect is noted with an increase in growth temperature. The observations are discussed from the standpoint of crystal growth theories.

1. Introduction

Crystal growth involves a number of processes that take place at the crystal-medium interface. Consequently, investigations of surface morphology of crystals serve as a useful indicator of the mechanism of crystal growth and the generation of defects. The literature on the surface morphology mainly deals with the type of sculptures observed on crystal surfaces and their connection with crystal growth theories (SUNAGAWA; SUNAGAWA, BENNEMA; VAN ENCKEVORT et al. 1980, 1981; SZURGOT, SANGWAL; SANGWAL et al.). Changes in the geometry of sculptures have been observed with a change in growth conditions (SUNAGAWA; SOKOŁOWSKI et al.; SZURGOT, SANGWAL). The quality of crystals is also connected with growth conditions (SZURGOT et al.; SANGWAL; SANGWAL et al.; LEFAUCHEUX et al.; GITS-LEON et al.). Therefore, it is interesting to correlate crystal quality with surface micromorphology under known experimental conditions. The present paper is devoted to such a study on potassium bichromate ($K_2Cr_2O_7$, KBC) crystals grown from aqueous solutions.

2. Experimental

For the investigation of surface morphology, KBC crystals were grown on seeds by decreasing the temperature of a saturated solution at a constant cooling rate. The stirring rate used was 16 cm/s. After growth the crystal was dried between filter papers and examined under an optical microscope in reflection. For the investigation of crystal quality, the crystals were grown by solvent evaporation method at a constant temperature, and examined in transmission. The details of the experimental procedures for growth and examination of the crystals are given elsewhere (SZURGOT et al.; SZURGOT, SANGWAL; SANGWAL, SZURGOT).

3. Results

3.1. Surface morphology

It was observed that the general surface relief of a face depends on supersaturation and temperature of growth. The effect is marked on {001} and {100} faces, which exhibit a pronounced anisotropy in the growth rates of parallel opposite faces. At low temperatures, the crystal surfaces are relatively smooth at low supersaturations, but they show thicker layers with increasing supersaturation. This feature is illustrated in Figure 1. It was also found that an increase in temperature leads to the development of relatively thin layers on a face under comparable supersaturation.

A change in bulk supersaturation at a particular temperature results into a change in the shape of macrospirals. This effect is significant for macrospirals on {001} faces, where the growth layers spreading along [$\bar{1}\bar{1}$0], [1$\bar{1}$0] and [010] directions have a tendency to become rounded (Fig. 1). On these faces the macrospirals also become elongated along [100] direction with an increase in supersaturation.

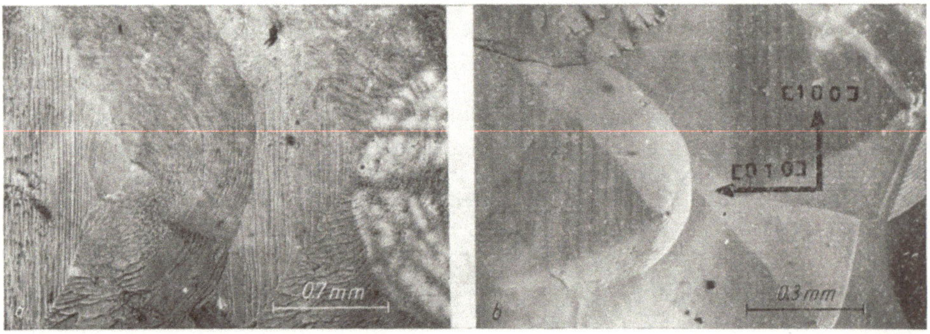

Fig. 1. Appearance of macrospirals on the (001) face of KBC obtained at the final temperature of 30 °C at cooling rates of (a) 0.4 and (b) 1.0 °C/h. At this temperature a cooling rate of 1 °C/h corresponds to about 4 % supersaturation

On the (001) face the angle α of intersection between layers moving along [010] and [$\bar{1}$00], and along [010] and [100] (Fig. 2) decreases to 20° from 150° with an increase in supersaturation at 37—25 °C, and can be expressed by the empirical relation

$$\operatorname{arc} \alpha = A(d\theta/dt)^n , \qquad (1)$$

where the constant $A = 1.387 \pm 0.052$ $(°C/h)^{1/2}$, $n = -0.50 \pm 0.04$, and the rate of cooling $d\theta/dt$ (°C/h) is a measure of supersaturation. The same dependence is obeyed by macrospirals on (00$\bar{1}$) face.

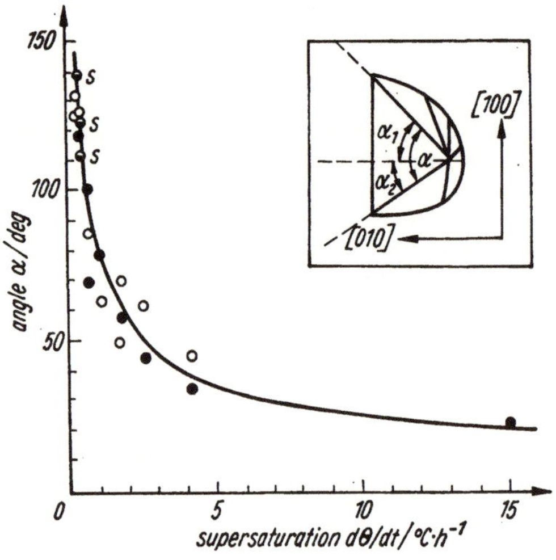

Fig. 2. Dependence of angle α for macrospirals produced on {001} faces on supersaturation $d\theta/dt$ in the temperature interval 37 —25 °C. The angles α, α_1 and α_2 for (001) face are shown in the inset. The filled and open circles denote the values for (001) and (00$\bar{1}$) faces, respectively. The points denoted by S are for macrospirals observed on crystals grown by solvent evaporation method but the supersaturation is unknown. A cooling rate of 1 °C/h corresponds to about 4 % supersaturation

3.2. Capture of inclusions at thick layers

The as-grown surfaces of KBC contain a large number of layers and steps and, in general, it is difficult to assess their role in the capture of inclusions and generation of defects during crystal growth. In order to investigate their role in the generation of defects, it is convenient to observe the process on cleaved surfaces, as shown by the growth patterns on the same region of a $(00\bar{1})$ cleavage face seen in Figure 3. It is obvious from the figure that inclusions appear at the thick cleavage step, which develops into macrokinks composed of equilibrium facets along [010] and [100] directions. As noted in Figure 1, some sides of the macrospirals are rounded. At relatively high supersaturations, inclusions are captured in the rounded parts of the macrospirals, as illustrated in Figure 4.

Fig. 3. Photographs illustrating the sequence of capture of inclusions I at a moving macrostep M on the $(00\bar{1})$ face. (a) Cleavage face, and (b, c) growth patterns. Growth was carried out at 28 °C and a supersaturation ≈ 12 % for (b) 30 s and (c) 5 min. The direction of movement of the macrostep is indicated by an arrow

Fig. 4. Anisotropy in the capture of inclusions observed during the development of a macrospiral on the $(00\bar{1})$ face of KBC at 23 °C. Supersaturation ≈ 10 %

3.3. Crystal quality

The crystal quality deteriorates with an increase in supersaturation, while it improves with an increase in temperature. This feature of the effect of these parameters may be noted from Figures 5a and b. It was also observed that the crystal quality improves (Fig. 5c) when the temperature stabilization is improved by using additional water jackets around the growth vessel. This feature may be noted from a comparison of the crystals seen in Figures 5b and 5c.

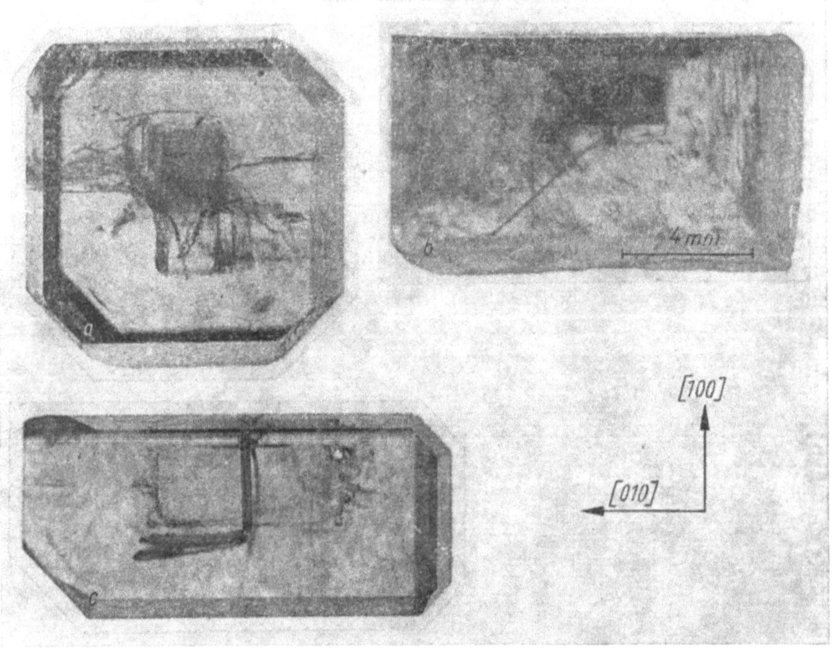

Fig. 5. KBC crystals grown by solvent evaporation method at (a) 55, (b) 30 and (c) 21 °C. Crystals (a) and (b) were obtained at a temperature control of about ±0.1 °C, whereas that in (c) was grown with a better temperature stabilization

4. Discussion

From the above results the following points may be noted:

1. The growth layers of macrospirals become thicker with an increase in supersaturation (Fig. 1), but an increase in temperature results into the development of relatively thin layers.
2. Inclusions are easily trapped at relatively thick layers (Fig. 3). Thick rounded layers are the preferred sites for the capture of inclusions (Fig. 4).
3. The angle α decreases with an increase in supersaturation (Fig. 2).
4. Crystal quality deteriorates with an increase in supersaturation but it improves with an increase in growth temperature (Fig. 5).

In order to explain Points 1, 2, and 4 we consider the bunching of growth layers. Fluctuations in growth conditions, high supersaturations and impurities present in solutions are the main causes of the formation of bunches (FRANK; CABRERA, VERMILYEA). The bunched layers transform into macrosteps which rearrange into equilibrium facets belonging to the crystallographic zone of faces of the crystal normal to the surface (Fig. 3) (RODRIGUEZ et al.; SANGWAL et al.). The macrokinks or teeth formed by the adjoining equilibrium facets act as sites for the capture of liquid inclusions (Figs. 3 and 4) and impurity aggregates, and for the generation of dislocations. Since defect content in crystals determine their quality, the crystal quality deteriorates with an increase in supersaturation.

The improvement in crystal quality with an increase in growth temperature may also be explained in terms of suppression of bunch formation. The suppression in bunch formation is due to the smoothening of the thicker layers by the enhanced mobility at high temperatures of relatively large chemical entities unfavourable for growth of substances with a positive temperature coefficient of solubility. Alternatively, it may be due to the annealing of growth layers at elevated temperatures.

Now we analyse the dependence of angle α on supersaturation. For this purpose we split α into α_1 and α_2, which are the angles between [010] and [100], and between [010] and [$\bar{1}$00], respectively (see Fig. 2). The values of A and n then are 1.077 and -0.38, and 0.456 and -0.73 for α_1 and α_2, respectively. For simplicity, we consider the supersaturation dependence of the rate v_∞ of displacement of a parallel array of straight steps with a large average spacing between them (OHARA, REID):

$$v_\infty = 2D_s c_0 \Omega \beta \Lambda \sigma / h x_s \,, \tag{2}$$

where the step retardation factor

$$\beta = \frac{1 - (\tau_D/\tau_S)}{1 + 2(\tau_D/\tau_S)\,(x_s/h)}\,, \tag{3}$$

the kink retardation factor Λ decreases strongly with an increase in the mean distance between kinks in a step and weakly depends on supersaturation σ, h is the height of a unit layer, D_s is the solute diffusion coefficient on the surface, c_0 is the equilibrium solute concentration on the surface, Ω is the molecular volume, x_s is the mean diffusion distance for the growth units, and τ_S and τ_D are the mean lifetimes for adsorbed growth units on a surface and a step, respectively. Since $x_s/h \approx 10^2-10^3$ and $\tau_D < \tau_S$ (OHARA, REID), it follows from eq. (3) that β is constant. Assuming that $\Lambda \propto x_0^{-m}$ (where m is a constant, and x_0 is the mean distance between kinks on a step), for Λ close to unity, we have

$$v_\infty = B\sigma / x_0^m x_s \,, \tag{4}$$

where B is a constant which is probably a function of x_0. The value of x_0 is expected to be different for steps along different directions on a face. Therefore, from eq. (4), for the approximation $\alpha \approx \tan \alpha$ with respect to steps 1 and 2, one obtains

$$\alpha = \frac{B_1 x_{02}^{m'} x_{s2}}{B_2 x_{01}^{m''} x_{s1}}\,, \tag{5}$$

where the exponents m' and m'' correspond to steps 1 and 2, respectively. It is difficult to establish the dependence of x_0 on supersaturation. Perhaps the simplest approach is to deduce it from the radii of stable semicircular two-dimensional nuclei formed ahead of steps. For such nuclei the excess free energy for their formation is $\Delta G^*/2$, but the radius is given by the expression for the radius r^* of critical two-dimensional circular nuclei i.e.

$$r^* = \gamma \Omega / kT\sigma \,. \tag{6}$$

Assuming that $r^* \propto x_0$ and $\sigma \propto d\theta/dt$, from eqs. (5) and (6) we obtain eq. (1).

The authors are grateful to Dr R. Rodriguez-Clemente for his constructive suggestions for the improvement of the discussion section. This work was carried out under Research Project CPBP 01.06. 6.04.

References

CABRERA, N., VERMILYEA, D. A.: in: Growth and Perfection of Crystals, eds R. H. DOREMUS, B. W. ROBERTS, D. TURNBULL, New York 1958, p. 393

FRANK, F. C.: in: Growth and Perfection of Crystals, eds. R. H. DOREMUS, B. W. ROBERTS, D. TURNBULL, New York 1958, p. 411

GITS-LEON, S., LEFAUCHEUX, F., ROBERT, M. C.: J. Cryst. Growth **44**, 345 (1978)

LEFAUCHEUX, F., ROBERT, M. C., SANGWAL, K.: J. Cryst. Growth **67**, 541 (1984)

OHARA, M., REID, R. C.: Modeling Crystal Growth Rates from Solution, New Jersey 1973

RODRIGUEZ, R., AQUILO, M., TEJADA, J.: J. Cryst. Growth **47**, 518 (1979)

SANGWAL, K.: Sci. Bull. Łódź Techn. Univ., Phys. Ser. in press

SANGWAL, K., RODRIGUEZ-CLEMENTE, R., VEINTEMILLAS-VERDAGUER, S.: J. Cryst. Growth **78**, 144 (1986)

SANGWAL, K., SZURGOT, M.: J. Cryst. Growth **80**, 351 (1987)

SOKOŁOWSKI, W., WOJCIECHOWSKI, K., SANGWAL, K., RODRIGUEZ-CLEMENTE, R.: Cryst. Res. Technol. **19**, 151 (1984)

SUNAGAWA, I.: Bull. Mineral. **104**, 81 (1981)

SUNAGAWA, I., BENNEMA, P.: in: Preparation and Properties of Solid State Materials, Vol. 7, ed. W. R. WILCOX, New York 1982, pp. 1–130

SZURGOT, M., SANGWAL, K.: J. Cryst. Growth **79**, 829 (1986)

SZURGOT, M., SANGWAL, K., KARNIEWICZ, J.: Cryst. Res. Technol. **20**, 645 (1985)

VAN ENCKEVORT, W. J. P., BENNEMA, P., VAN DER LINDEN, W. H.: Z. Phys. Chem. N.F. **134**, 171 (1981)

VAN ENCKEVORT, W. J. P., JANSSEN-VAN ROSMALEN, R., VAN DER LINDEN, W. H.: J. Cryst. Growth **409**, 502 (1980)

(Received April 4, 1987)

Authors' addresses:

Dr. M. SZURGOT, Doc. Dr. hab. K. SANGWAL*

Institute of Physics
Technical University of Łódź
Wólczańska 219
93005 Łódź
Poland

*) Permanent address:

Department of Physical and Solid State Chemistry
Pedagogical University
Al. Zawadzkiego 13/15
42200 Częstochowa
Poland

Cryst. Res. Technol.	22	1987	11	1403–1408

D. Scharfenberg-Pfeiffer, E. Höhne, M. Wunderwald

Zentralinstitut für Molekularbiologie der Akademie der Wissenschaften der DDR, Berlin-Buch and
Zentralinstitut für Mikrobiologie und Experimentelle Therapie der Akademie der Wissenschaften der DDR, Jena

Crystal- and Molecular Structure of 19-nor-Digitoxigenin

19-nor-Digitoxigenin crystallizes in the orthorhombic space group P $2_12_12_1$ with four formula units $C_{22}H_{32}O_4$ per unit cell. The lattice parameters are $a = 7.082$, $b = 12.476$, $c = 21.486$ Å. The crystal structure was determined by direct methods and refined by a least-squares procedure to the discrepancy factor $R = 0.042$. – The orientation of the lactone ring relative to the steroid skeleton was found to be the opposite of that known for digitoxigenin.

19-nor-Digitoxigenin kristallisiert in der orthorhombischen Raumgruppe P $2_12_12_1$ mit vier Formeleinheiten $C_{22}H_{32}O_4$ in der Elementarzelle. Die Gitterkonstanten sind $a = 7{,}082$, $b = 12{,}476$, $c = 21{,}486$ Å. Die Kristallstruktur wurde mit direkten Methoden bestimmt und mit der Methode der kleinsten Quadrate bis zu einem R-Wert von 0.042 verfeinert. – Die gefundene Orientierung des Laktonrings relativ zum Steroidgrundkörper ist entgegengesetzt der, die man von Digitoxigenin kennt.

Introduction

Within a research program on cardiotonic steroids 19-nor-digitoxigenin was synthesized by total synthesis (Wunderwald et al.). Its pharmacological activity was investigated by in vitro and in vivo tests. The structure of the new compound was of interest in connection with structure-activity-relationships. In this paper the crystal and molecular structure of 19-nor-digitoxigenin is described.

Crystal data

Single crystal diffractometry, graphite monochromated MoKα radiation, $\lambda = 0.71069$ Å, orthorhombic space group P $2_12_12_1$, $a = 7.082(1)$, $b = 12.476(2)$, $c = 21.486(3)$ Å, number of formula units $C_{22}H_{32}O_4$ in the unit cell $Z = 4$, calculated density $D_x = 1.261$ g cm^{-3}, crystal size $0.6 \times 0.4 \times 0.4$ mm^3.

Intensity data collection, structure determination and refinement

The intensities of 1937 independent reflections were collected in the range $3° < 2\theta < 50°$ on an ENRAF-NONIUS CAD-4 diffractometer, using MoKα radiation and the $\omega - 2\theta$ scan technique. 1729 reflections had net intensities $I(hkl) > 2\sigma(I)$ and these were considered as observed in the structure analysis. Corrections for Lorentz and polarization effects were carried out but no absorption correction was made.

A MULTAN 82 (MAIN et al.) procedure, in which 3 special and 4 general E-values additionally to the origin defining reflections were used, yielded the positions of all non-hydrogen atoms. The blocked full-matrix least-squares refinement was carried out with SHELX-76 (SHELDRICK), using unit weights. All hydroxgens were located in a difference Fourier synthesis but, after refinement of their positions, most of them had to be rejected because of unreliable C–H bond length. Thus, all hydrogen positions, except those of the hydroxyl protons, were calculated and constrained (riding model) during the last two cycles of anisotropic refinement. Convergence was achieved at a conventional R-value of 0.042.

The resulting coordinates and their e.s.d.'s are given in Table 1.

Table 1

Fractional atomic coordinates

Atom	$x/a(\sigma)$	$y/b(\sigma)$	$z/c(\sigma)$	Atom	$x/a(\sigma)$	$y/b(\sigma)$	$z/c(\sigma)$
C1	4044(6)	5723(3)	5995(2)	H22	352	666	518
C2	2874(6)	5989(3)	5414(2)	H31	−1	633	516
C3	839(6)	6280(3)	5580(2)	H41	−143	568	613
C4	−40(6)	5412(3)	5990(2)	H42	−16	468	572
C5	1122(6)	5170(3)	6568(2)	H51	119	590	684
C6	181(6)	4335(3)	6986(2)	H61	−129	454	704
C7	308(5)	3206(3)	6718(2)	H62	86	435	744
C8	2370(5)	2910(3)	6587(2)	H71	−27	265	705
C9	3251(5)	3696(3)	6118(2)	H72	−50	316	629
C10	3168(5)	4840(3)	6395(2)	H81	318	300	701
C11	5281(5)	3370(3)	5970(2)	H91	248	368	568
C12	5397(5)	2199(3)	5756(2)	H101	402	478	681
C13	4640(5)	1400(3)	6242(2)	H111	582	388	561
C14	2573(5)	1721(3)	6414(2)	H112	613	346	638
C15	1417(5)	1338(3)	5859(2)	H121	457	212	534
C16	2280(6)	253(3)	5687(2)	H122	685	201	566
C17	4386(6)	290(3)	5897(2)	H151	−5	124	598
C18	5974(5)	1349(3)	6800(2)	H152	153	190	548
C20	4984(6)	−668(3)	6269(1)	H161	154	−39	592
C21	3768(6)	−1198(3)	6751(2)	H162	221	14	519
C22	6624(6)	−1178(3)	6239(2)	H171	534	24	551
C23	6646(7)	−2032(3)	6705(2)	H181	547	84	717
O14	1958(4)	1074(2)	6935(1)	H182	610	216	697
O21	4979(4)	−2027(2)	7007(1)	H183	734	107	664
O23	7871(5)	−2673(2)	6838(1)	H211	250	−153	655
O3	841(5)	7297(2)	5904(1)	H212	340	−63	712
H11	542	546	585	H221	777	−99	592
H12	419	644	627	HO14	209	144	734
H21	287	530	511	HO3	7	738	624

Discussion

The molecular structure of 19-nor-digitoxigenin is depicted in Figure 1. The orientation of the lactone ring relative to the steroid skeleton, as given by the torsion angle C13—C17—C20—C22, is −97.2°. For digitoxigenin, however, 76.3° were observed by KARLE and KARLE. Both values fall into separate domains of theoretical energy minima of free cardenolide molecules, as calculated quantum chemically (SCHARFENBERG, SKLENAR; PFEIFFER et al., 1986) by the CNDO/2 method. It is interesting to note the dependence of the bond angles C17—C20—C22, C17—C20—C21 on the lactone

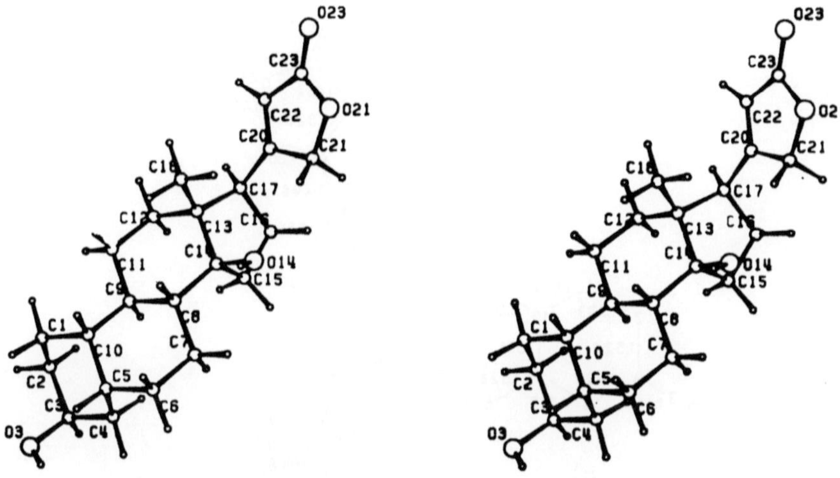

Fig. 1. Stereoview of the molecule

Table 2

Lactone ring orientation and bond angle variations, as observed in crystal structures of cardenolides (I = Tors. ∢ C13—C17—C20—C22; II = ∢ C17—C20—C22; III = ∢ C17—C20—C21; all values in deg.)

Compound	I	II	III	Compound	I	II	III
a)	73.6	130.6	120.0	g)	−109.2	125.4	124.2
b)	90.6	135.2	117.5	h)	−98.9	126.4	125.5
c)	65.8	132.9	117.3	i)	−95.7	126.2	125.9
d)	52.3	133.5	119.4	j)	−123.0	125.0	125.7
e)	65.7	132.0	118.0	k)	−94.2	126.4	126.0
f)	65.0	132.6	118.5	l)	−128.7	125.1	127.3
				m)	−110.9	124.3	124.4
				n)	−97.2	127.2	123.8
mean value	68.8	132.8	118.4	mean value	−107.2	125.8	125.4

a) digitoxigenin (KARLE, KARLE)
b) 12β-acetyl-digitoxigenin (MESSERSCHMIDT, HÖHNE, STRECKENBACH)
c) 3-epi-digitoxigenin (MESSERSCHMIDT, HÖHNE, MEGGES 1981 a)
d) 5β-hydroxy-gitoxigenin (PRZYBYLSKA, AHMED)
e) 3β,16β-di-acetyl-gitoxigenin (RECK et al.)
f) oleandrin (KARTHA, GO)
g) 3β-acetyl-digitoxigenin (MESSERSCHMIDT, HÖHNE, MEGGES 1981 b)
h) digoxin (GO et al.)
i) digoxigenin monodigitoxoside (GO, KARTHA 1982 a)
j) ouabain diethanol (GO, KARTHA 1981)
k) digoxigenin bisdigitoxoside (GO, KARTHA 1982 b)
l) gitoxin (GO, KARTHA 1980)
m) strophanthidin (molecula A) (GILARDI, FLIPPEN)
n) 19-nor-digitoxigenin (title compound)

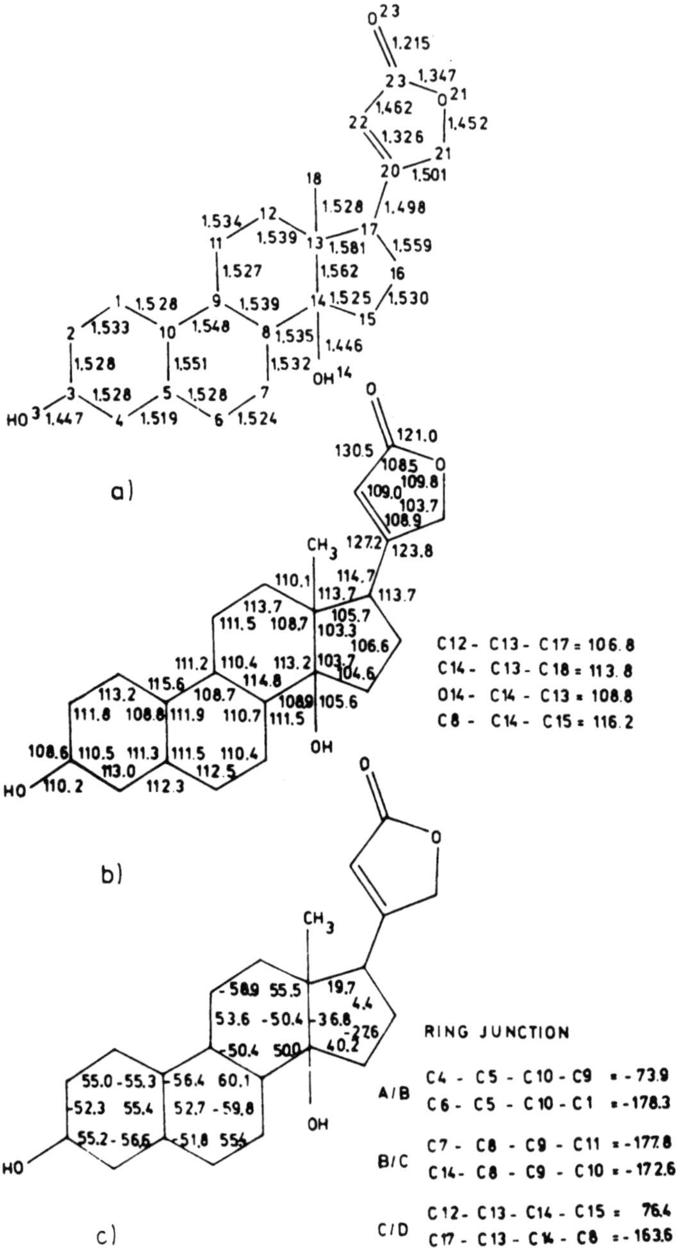

Fig. 2. Intramolecular dimensions of 19-nor-digitoxigenin. a) bond distances (Å); $\sigma_{range} = 0.005 - 0.008$ Å; b) bond angles (°); $\sigma_{range} = 0.3 - 0.5$°; c) endocyclic torsion angles (°); $\sigma_{range} = 0.6 - 0.9$°

ring orientation (GO, KARTHA; HÖHNE, PFEIFFER). Numerical data are compiled in Table 2.

Figure 2 shows the bond distances, bond angles and selected torsion angles of the title compound. They do not markedly differ from those of other cardenolides.

For comparisons, the 19-nor-digitoxigenin molecule was superimposed (program FITMO; KRETSCHMER) with digitoxigenin (KARLE, KARLE) and (20R)-19-nor-5β,14β-pregnane-3β,14,20-triol 3,20-diacetate (PFEIFFER et al. 1987) by a least squares fit of the atomic positions of C1 to C14, resulting in root mean square separations of 0.09 Å and 0.05 Å, respectively. The latter applies to both symmetry independent molecules described in the afroementioned study.

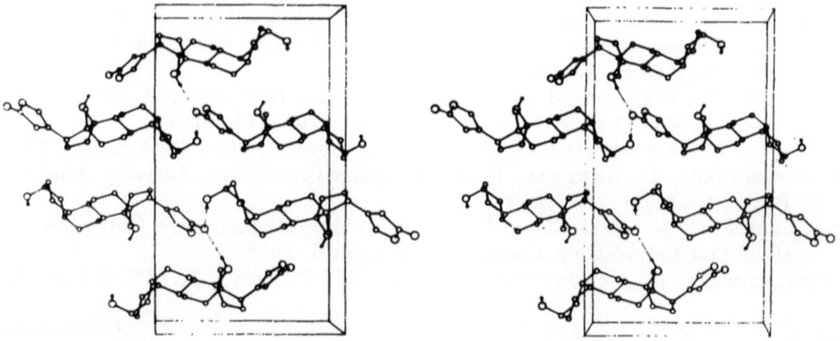

Fig. 3. Molecular packing in 19-nor-digitoxigenin viewed along *a*

The analysis of the ring conformations was carried out with the program RING (PÁRKÁNYI). The rings A, B, and C of the title compound adopt a chair conformation with ω-values of 55.0, 56.0, and 53.1°, respectively, comparable with the 55.9° reported for free cyclohexane (GEISE). The conformation of the five membered D-ring $\Delta C_s^{C14} = 6.1°$ and $\Delta C_2^{C17} = 12.7°$ (for definition see DUAX et al.) is intermediate between C14β envelope and C14β, C15α half-chair, whereas the five membered lactone ring has the expected planarity. Figure 3 shows the arrangement of the molecules of 19-nor-digitoxigenin in the unit cell, viewed along the *a*-axis. The molecules are linked head to tail by the hydrogen bond O23 ... HO3′$(1 + x, y - 1, z)$ = 2.02 Å, thus forming infinite chains. Another hydrogen bond, viz. O23 ... HO14′$(1 - x, y - 0.5, 1.5 - z)$ = = 2.08 Å, connects these chains to a three dimensional network (O23 ... O3′) = 2.91 Å, O23 ... O14′ = 3.07 Å).

References

DUAX, W. L., WEEKS, C. M., ROHRER, D. C.: Topics in Stereochemistry, Vol. 9, edited by E. L. ELIEL & ALLINGER, New York 1976, pp. 271–383

GEISE, H. J., BUYS, H. R., MIJLHOFF, F. C.: J. Mol. Struct. **9**, 447 (1971)

GILARDI, R. D., FLIPPEN, J. L.: Acta Cryst. **B29**, 1842 (1973)

GO, K., KARTHA, G.: Acta Cryst. **B36**, 3034 (1980)

GO, K., KARTHA, G.: Cryst. Struct. Comm. **10**, 1329 (1981)

GO, K., KARTHA, G.: Cryst. Struct. Comm. **11**, 279 (1982) a

GO, K., KARTHA, G.: Cryst. Struct. Comm. **11**, 285 (1982) b

GO, K., KARTHA, G., CHEN, P. J.: Acta Cryst. **B36**, 1811 (1980)

HÖHNE, E., PFEIFFER, D.: studia biophysica **97**, 81 (1983)

KARLE, I. L., KARLE, J.: Acta Cryst. **B25**, 434 (1969)

KARTHA, G., GO, K.: Cryst. Struct. Comm. **10**, 1323 (1981)

KRETSCHMER, R.-G.: FITMO. A Program to Superimpose two Molecules by Least Squares Procedure. Zentralinstitut für Molekularbiologie der Akademie der Wissenschaften der DDR, Berlin-Buch 1984

MAIN, P., FISKE, S. J., HULL, S. E., LESSINGER, L., GERMAIN, G., DECLERCQ, J.-P., WOOLFSON, M. M.: MULTAN 82. A System of Computer Programs for the Automatic Solution of Crystal Structures from X-ray Diffraction Data. Univs of York, England and Louvain, Belgium 1982
MESSERSCHMIDT, A., HÖHNE, E., STRECKENBACH, B.: Cryst. Struct. Comm. **10**, 141 (1981)
MESSERSCHMIDT, A., HÖHNE, E., MEGGES, R.: Cryst. Struct. Comm. **10**, 149 (1981) a
MESSERSCHMIDT, A., HÖHNE, E., MEGGES, R.: Cryst. Struct. Comm. **10**, 157 (1981) b
PÁRKÁNYI, L.: RING. A Program of Conformation Analysis, Budapest 1979
PFEIFFER, D., HÖHNE, E., SCHARFENBERG, P.: Crystallographic Computing. Proceedings of the International Summer School, edited by P. PAUFLER, V. GNEIST & D. KLIMM, Karl-Marx-Univ. Leipzig 1986, p. 231
PFEIFFER, D., WEILAND, J., LEIBNITZ, P.: Cryst. Res. Technol. (1987) **22**, 389 (1987)
PRZYBYLSKA, M., AHMED, F. R.: Acta Cryst. **B35**, 2436 (1979)
RECK, G., MESSERSCHMIDT, A., MEGGES, R.: Cryst. Struct. Comm. **10**, 637 (1981)
SCHARFENBERG, P., SKLENAR, H.: Quantitative Structure-Activity Analysis (Suhl 1976), Berlin 1978, pp. 175–189
SHELDRICK, G. M.: SHELX. A Program for Crystal Structure Determination, University Chemical Laboratory, Cambridge, England, 1976
WUNDERWALD, M., SCHUBERT, G., EIBISCH, H., PONSOLD, K.: DDR Pat. 236737 (1986)

(Received April 4, 1987)

Authors adresses:

Dr. Doris SCHARFENBERG, Prof. Dr. habil. E. HÖHNE
Akademie der Wissenschaften der DDR
Zentralinstitut für Molekularbiologie
Robert-Rössle-Str. 10
Berlin-Buch
DDR-1115

Dr. Manfred WUNDERWALD
Akademie der Wissenschaften der DDR
Zentralinstitut für Mikrobiologie und
Experimentelle Therapie
Beutenbergstr. 11
Jena
DDR-6900

Cryst. Res. Technol.	22	1987	11	1409–1414

N. S. Orlova, I. V. Bodnar

Institute of Physics of Solids and Semiconductors
Byelorussian Academy of Sciences, Minsk, USSR

Thermal Expansion in $CuAlTe_2$ Compound

The lattice parameters a and c as well as the axial thermal expansion coefficients $\alpha_\perp$ and $\alpha_\parallel$ in the $CuAlTe_2$ chalcopyrite-type compound are determined as a function of temperature in the range from 80 to 650 K by a X-ray diffractometry technique. The data obtained are used to evaluate the axial ratio c/a, the tetragonal distortion $\delta = 2 - c/a$, the interatomic distances for Cu–Te and Al–Te bonds and their temperature coefficients. It is found that the thermal expansion behaviour of $CuAlTe_2$ is similar to that of other $CuB^{III}C_2^{VI}$ compounds in having a relatively small expansivity along the c-axis and a large one in the perpendicular direction. When comparing the results for a series of the $CuB^{III}C_2^{VI}$ compounds (B–Al, Ga, In; C–S, Se, Te) it is shown that the correlations between the thermal expansion coefficients $\alpha_\perp$, $\alpha_\parallel$, α_m, $d\delta/dT$ and the tetragonal distortion δ, as well as the molar mass of the compound take place.

Рентгенографическим методом определены температурные зависимости параметров решетки и главных коэффициентов теплового расширения соединения $CuAlTe_2$ со структурой халькопирита в интервале температур 80—650 К. На основании полученных данных найдено изменение с температурой осевого отношения c/a, параметра тетрагонального искажения $\delta = 2 - c/a$, длин связи между атомами Cu–Te и Al–Te и их термических коэффициентов. Установлено, что в $CuAlTe_2$, так же как и в других соединениях $CuB^{III}C_2^{VI}$, тепловое расширение анизотропно с коэффициентом расширения вдоль направления, параллельного тетрагональной оси, гораздо меньшим, чем вдоль перпендикулярного ей направления. Сравнительный анализ результатов для девяти соединений $CuB^{III}C_2^{VI}$ (B–Al, Ga, In; C–S, Se, Te) показал существование корреляций коэффициентов теплового расширения и его анизотропии с параметром тетрагонального искажения и мольной массой соединения.

1. Introduction

In the course of a systematic investigation of the lattice thermal behaviour of the ternary $A^IB^{III}C_2^{VI}$ type semiconductors with the chalcopyrite structure in this work we have studied by a X-ray diffraction method the temperature dependence of the lattice parameters, a and c, that of the axial ratio c/a, the tetragonal distortion $\delta = 2 - c/a$ as well as their thermal expansion coefficients for the $CuAlTe_2$ compound. The experimental data obtained are compared with the results of our previous measurements for other eight $CuB^{III}C_2^{VI}$ chalcopyrite-type compounds (Bodnar, Orlova) and the variation of these structure parameters and their temperature coefficients while substituting the Al–Ga–In cations and the S–Se–Te anions is discussed.

2. Experimental procedure

The apparatus and the experimental procedure used in determining the lattice parameters and their thermal expansion coefficients were the same as those described earlier (Bodnar, Orlova). Measurements of the diffraction line angle sites were made using a standard diffractometer with a scintillation-counter and the crystal-monochromatized CuKα radiation in an X-ray temperature vacuum camera (SIROTA et al.; SIROTA, ZHABKO) at 80, 150, 220, 290, 360, 430, 500, 570, 650 K. The temperature could be controlled within ±2 K.

These studies were carried out on the $CuAlTe_2$ samples obtained by one-temperature technique with vibration. Chemical, X-ray, microstructure, and differential thermal analyses carried out by us showed the specimens to be homogeneous, single phase, and uniform. For X-ray powder diffraction work the crystal samples were crushed to a powder of appropriate particle size. To remove strain the powder samples were annealed at about 600 K during 12 h in vacuum. The angle positions of 19 diffraction lines with different Miller indexes hkl were used to determine the lattice parameters a and c at each temperature, employing Cohen's least-squares method (MIRKIN).

3. Results

The lattice parameters obtained at different temperatures have been employed to calculate the effect of temperature on the axial ratio c/a, the tetragonal distortion of the lattice δ. Moreover using the Abrahams-Bernstein relations and taking account the sublattice distortion ($\Delta y = -0.0008$) we have estimated the positional parameter x, which denotes the location of the Te atom relative to the Cu and Al atoms, as well as the interatomic distances l_{AC} and l_{BC} for the Cu–Te and Al–Te bonds, respectively.

The values of these parameters obtained at 300 K are $a = 0.602307$ nm, $c = 1.192332$ nm, $c/a = 1.9796$, $\delta = 0.0204$, $x = 0.2543$, $l_{AC} = 0.26144$ nm, $l_{BC} = 0.25842$ nm. They are slightly different from the available literature data (ABRAHAMS, BERNSTEIN). The discrepancy between the results can be due to the difference in the defect concentration of the investigated samples.

The variation of the all structure parameters with temperature was described by the polynominal of n-th order in temperature as follows

$$a = 0.600818 + 3.26 \cdot 10^{-6}T + 6.09 \cdot 10^{-9}T^2 - 1.45 \cdot 10^{-12}T^3, \text{ nm}$$
$$c = 1.190740. + 4.24 \cdot 10^{-5}T + 3.33 \cdot 10^{-9}T^2 + 1.91 \cdot 10^{-12}T^3, \text{ nm}$$
$$c/a = 1.9819 - 3.70 \cdot 10^{-6}T - 1.45 \cdot 10^{-8}T^2 + 8.08 \cdot 10^{-12}T^3,$$
$$\delta = 0.0181 + 3.70 \cdot 10^{-6}T + 1.45 \cdot 10^{-8}T^2 - 8.08 \cdot 10^{-12}T^3,$$
$$x = 0.2538 + 9.34 \cdot 10^{-9}T + 3.66 \cdot 10^{-9}T^2 - 2.03 \cdot 10^{-12}T^3.$$

By differentiating these polynomials ($d \ln P(T)/dT$) we determined the coefficients of the thermal expansion parallel, $\alpha_{\parallel}$, and perpendicular, α_{AC}, to the principal c-axis, the thermal expansion coefficients along the A–C, α_{AC}, and the B–C, α_{BC}, bonds as well as the temperature coefficients of the parameter x, α_x, those of the tetragonal distortion δ, α_δ. The coefficients of the axial ratio c/a, $\alpha_{c/a} = \alpha_{\parallel} - \alpha_{\perp}$, and the derivation $d\delta/dT = (-c/a)(\alpha_{\parallel} - \alpha_{\perp})$ were also calculated. At room temperature these values are $\alpha_{\parallel} = 5.67$, $\alpha_{\perp} = 10.86$, $\alpha_m = (2\alpha_{\perp} + \alpha_{\parallel})/3 = 9.13$, $\alpha_{c/a} = -5.19$, $\alpha_{AC} = 12.67$, $\alpha_{BC} = 5.70$, $\alpha_x = 10.18$, $d\delta/dT = 10.00$, (in 10^{-6} K^{-1}); $\gamma_{\parallel}/\alpha_{\perp} = 0.53$, $A_\alpha = (\alpha_{\parallel} - \alpha_{\perp})/\alpha_{\perp} = -0.48$. Figure 1 shows the variation of all the coefficients of thermal expansion with temperature.

As seen for $CuAlTe_2$, as in the case of other $A^IB^{III}C_2^{VI}$ compounds with the chalcopyrite structure (BODNAR, ORLOVA; BRÜHL, NEUMANN et al. 1980, 1981; KISTAIAH, MURTHY; REDDY et al.) the coefficient of expansion along the c-axis, $\alpha_{\parallel}$, is smaller than that along the a-axis, $\alpha_{\perp}$. The coefficient α_{BC} for the more covalent bond B–C coincides with $\alpha_{\parallel}$ while the coefficient α_{AC} for the more ionic bond A–C is larger than the coefficient $\alpha_{\perp}$. The axial ratio decreases with increasing temperature, i.e. the distortion increases. Considering the temperature dependence of $\alpha_{\parallel}$ and $\alpha_{\perp}$ (see Fig. 1)

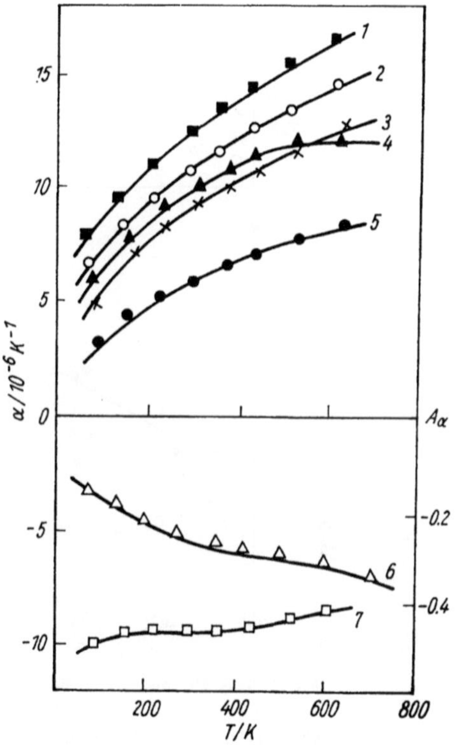

Fig. 1. Temperature dependence of the thermal expansion coefficients α_{AC} (1), $\alpha_\perp$ (2), α_m (3) $d\delta/dT$ (4), $\alpha_{||}$ and α_{BC}(5), $\alpha_{c/a} = \alpha_{||} - \alpha_\perp$ (6), the parameter $A_\alpha = (\alpha_{||} - \alpha_\perp)/\alpha_\perp$ (7) for the $CuAlTe_2$ compound

it seems to be evident that $\alpha_\perp$ increases more rapidly with increasing temperature than $\alpha_{||}$. In consequence of this behaviour the absolute values of $c_{c/a}$ and $d\delta/dT$ increase with temperature.

4. Discussion

As it is known the anisotropy of the thermal expansion coefficients for the uniaxial tetragonal structure can be determined by means of the temperature coefficient of the axial ratio $\alpha_{c/a} = \gamma_{||} - \alpha_\perp$ and the magnitude of this anisotropy can be evaluated by the ratio $A_{\alpha 1} = \alpha_{||}/\alpha_\perp$ or the parameter $A_\alpha = (\alpha_{||} - \alpha_\perp)/\alpha_\perp$ (at $\alpha_\perp > \alpha_{||}$) (Novikova). Moreover for the crystals with the chalcopyrite structure the anisotropy of the thermal expansion can be characterized by the derivative of the tetragonal distortion $d\delta/dT$ (Neumann) evaluated from the linear thermal coefficients $\alpha_{||}$ and $\alpha_\perp$ via the relation $d\delta/dT = (-c/a)(\alpha_{||} - \alpha_\perp)$.

The tetragonal distortion δ in the chalcopyrite structure has been one main parameter determining the structure anisotropy. Therefore it is interesting to examine the dependence of the thermal expansion coefficients and that of the derivative $d\delta/dT$ on the tetragonal distortion δ for all investigated $CuB^{III}C_2^{VI}$ compounds. Figure 2 shows the plot of α_{AC}, $\alpha_\perp$, α_m, $\alpha_{||}$, $d\delta/dT$, and $A_\alpha = (\alpha_{||} - \alpha_\perp)/\alpha_\perp$ against δ for nine $CuB^{III}C_2^{VI}$. The compounds are numbered in the sequence of the molar mass increase from $CuAlS_2$ (1) to $CuInTe_2$ (9). In this Figure the shaded symbols represent our results while the open symbols — literature data (Brühl et al.; Kistaiah, Murthy; Reddy et al.). The difference in the $\alpha_{||}$, $\alpha_\perp$, $d\delta/dT$ values reported by various authors can be ascribed to some systematic errors in both the experimental measurements

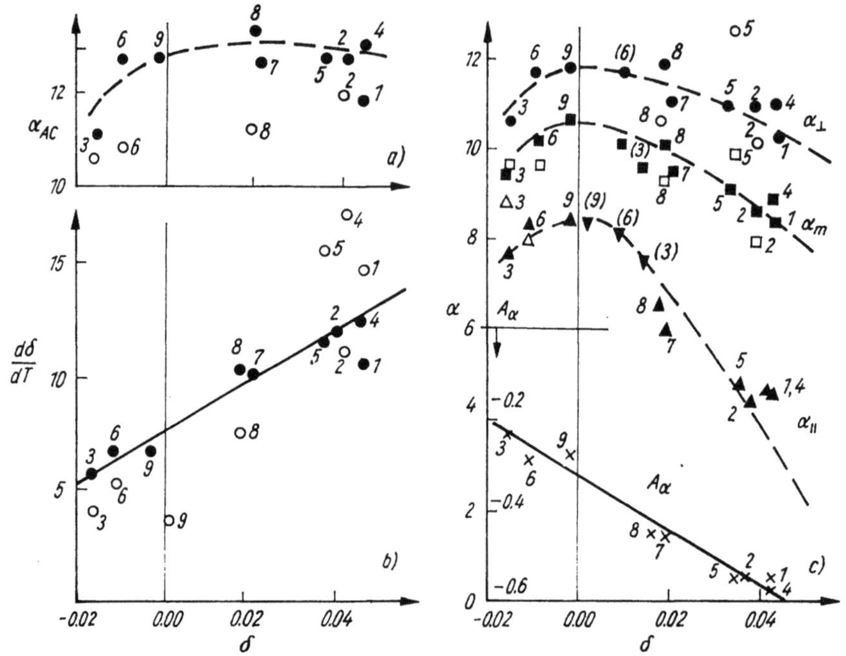

Fig. 2. The thermal expansion coefficients a) α_{AC} (●, ○), b) $d\delta/dT$ (●, ○), c) $\alpha_\perp$ (●, ○) α_m (■, □), $\alpha_{||}$ (▲, △) in 10^{-6} K^{-1} and $A_\alpha(x)$ as a function of the tetragonal distortion δ for the $CuAlS_2$ (1), $CuGaS_2$ (2), $CuInS_2$ (3), $CuAlSe_2$ (4), $CuGaSe_2$ (5), $CuInSe_2$ (6), $CuAlTe_2$ (7), $CuGaTe_2$ (8), $CuInTe_2$ (9) compounds. ●, ■, ▲ – our results, ○, □, △ – literature data

and the calculation procedures in deriving the lattice parameters and thermal expansion coefficients. To avoide this discrepancy while comparing the thermal expansion coefficients for different $CuB^{III}C_2^{VI}$ compounds we used only our results (BODNAR, ORLOVA) obtained under similar experimental conditions and the same data processing technique.

It is can be seen from Figure 2 that the experimental values of $d\delta/dT$ obtained by us for a given compound family is a linearly increasing function in δ which can be described, as expected from free-energy consideration in harmonic approximation (NOOLANDI; BRÜHL et al.; NEUMANN; BHAR, SAMANTA), by the following expression

$$d\delta/dT = C_0 + C_1\delta .$$

This relation can be obtained from the expansion of the free energy in terms of the tetragonal distortion δ with neglect of all terms of the third order and higher. The linear dependence of $d\delta/dT$ on δ shows that this harmonic approximation could be valid for the $CuB^{III}C_2^{VI}$ compounds considered. The parameters $C_0 = 7.6 \cdot 10^{-6}\ K^{-1}$ and $C_1 = 1.3 \cdot 10^{-4}\ K^{-1}$ estimated by least-square method are close to the values reported by NEUMANN (1981).

Furthermore, we see from Figure 2, that the absolute value of the anisotropy parameter A_α also rises linearly with the increasing tetragonal distortion δ. The AC-bond thermal expansion coefficient α_{AC} changes slightly with δ whilst the coefficients $\alpha_{||}$, $\alpha_\perp$, α_m monotonically decrease when the distortion δ increases. The most marked change is noted for the coefficient of expansion parallel to the tetragonal axis $\alpha_{||}$ which may become negative at $\delta > 0.7$.

It is interesting to note that the lowest anisotropy of the thermal expansion coefficients which is determined by the derivative $d\delta/dT$ or the parameter A_α is observed for the $CuInS_2$ compound. This anisotropy increases with rising δ which is mainly due to changes in $\alpha_{||}$. The largest values of the thermal expansion coefficients $\alpha_{||}, \alpha_\perp, \alpha_m$ are noted for $CuInTe_2$ for which the tetragonal distortion is close to zero.

Figure 3 shows the lattice parameters a and c, the molar volume V_M, the mean (α_m) and the two principal thermal expansion coefficients ($\alpha_\perp, \alpha_{||}$) as well as the tetragonal distortion δ and its derivative $d\delta/dT$ against molar mass M for all nine investigated $CuB^{III}C_2^{VI}$ compounds. This Figure also gives the melting temperatures T_m obtained by us and the values of $(M\Theta_D^2)^{1/2}$ evaluated from these data (here Θ_D is Debye characteristic temperature). As seen the molar volume V_M, the parameters a and c, the coefficients of thermal expansion $\alpha_{||}, \alpha_\perp, \alpha_m$ rise whilst the melting temperature T_m and the value of $(M\Theta_D^2)^{1/2}$ decrease with the increasing atomic number of the anion C (S → Se → Te) or the cation B (Al → Ga → In). This gives evidence for the decrease of the interatomic interaction with the molar mass increasing.

It is to be noted that the monotonical dependence of these structural and thermal parameters on the molar mass has not been observed. We can give two separate correlations for the $CuGaC_2^{VI}$ or $CuAlC_2^{VI}$ and $CuInC_2^{VI}$ compound series, respectively. The values of the lattice parameters a and c as well as the coefficients $\alpha_\perp$ and $\alpha_{||}$ are

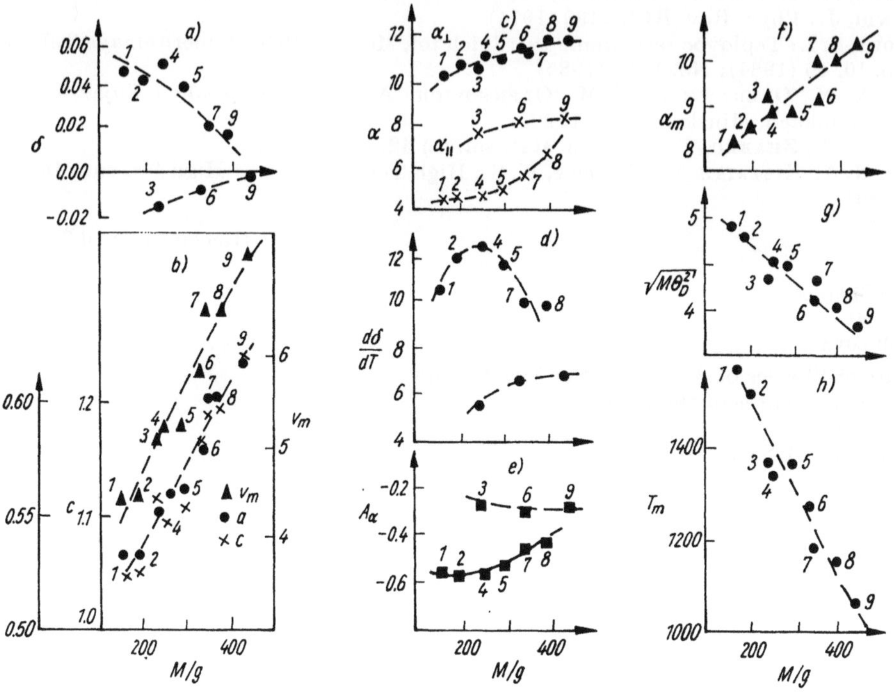

Fig. 3. Variation of a) the tetragonal distortion δ; b) the molar volume v_m (in 10^8 m³/moled), the lattice parameters a and c (in nm); c) the principal thermal expansion coefficients $\alpha_\perp$ and $\alpha_{||}$ (in 10^{-6} K^{-1}); d) the derivative $d\delta/dT$ (in 10^{-6} K^{-1}); e) the anisotropy parameter A_α; f) the mean linear thermal expansion coefficient $\alpha_m = (2\alpha_\perp + \alpha_{||})/3$ (in 10^{-6} K^{-1}); g) the value of $\sqrt{M\theta_D^2}$ (in 10^3 g K²); h) the melting temperature (in K) with the molar mass M (in g) for $CuAlS_2$ (1), $CuGaS_2$ (2), $CuInS_2$ (3), $CuAlSe_2$ (4), $CuGaSe_2$ (5), $CuInSe_2$ (6), $CuAlTe_2$ (7), $CuGaTe_2$ (8), $CuInTe_2$ (9)

close to the corresponding parameters and coefficients while substituting the Al atom (compounds 1, 4, 7) by the Ga atom (2, 5, 8). When these cations are substituated by the In atom (compounds 3, 6, 9) the values of $\alpha_\perp$ increase less than those of $\alpha_\parallel$. As a result the anisotropy of thermal expansion for the $CuInC_2^{VI}$ compounds decreases. In the case of anion substitution in the $CuAlC_2^{VI}$ and $CuGaC_2^{VI}$ compounds the tetragonal distortion δ and its temperature coefficient $d\delta/dT$ monotonically decrease as the atomic mass of chalcogen atom increases. For the $CuInC_2^{VI}$ compounds the values of δ and $d\delta/dT$ are smaller than ones for $CuAlC_2^{VI}$ and $CuGaC_2^{VI}$ and they almost are unchanged while sustituting anion C for heavier element in the S–Se–Te series.

References

ABRAHAMS, G. S. C., BERNSTEIN, J. L.: J. Chem. Phys. **59**, 5415 (1973)

BODNAR, I. V., ORLOVA, N. S.: phys. stat. sol. (a) **78**, K59 (1983); Izv. AN SSSR. Neorganich. mater. **21**, 1110 (1985); Cryst. Res. Technol. **21**. 1091 (1986)

BHAR, G. C., SAMANTA, L. K.: Phys. Rev. **B26**, 2275 (1982)

BRÜHL, H. G., NEUMANN, H., KÜHN, G.: Solid State Commun. **24**, 224 (1980)

BRÜHL, H. G., NEUMANN, H., PFEIFFER, T., KÜHN, G.: phys. stat. sol. (a) **66**, 597 (1981)

KISTAIAH, P., MURTHY, K. S.: J. Less-Common Metals **105**, 37 (1985)

MIRKIN, L. I.: Spravochnik po rentgenostrukturnimu analisu polikristallov, Moscow 1961

NEUMANN, H.: Kristall u. Technik **15**, 849 (1980); Cryst. Res. Technol. **18**, 659 (1983)

NOOLANDI, J.: Phys. Rev. **B10**, 2490 (1974)

NOVIKOVA, S. I.: Teplovoe rasshirenie tverdyich tel, Moscow 1974; Izmeritelnaya Teknika No. 10, 40 (1984); No. 1, 48 (1985)

SIROTA, N. N., OLEKNOVICH, N. M., OLEKNOVICH, A. I.: Chimicheskaya Svjaz v puloprovodnikach, Minsk 1969, p. 172

SIROTA, N. N., ZHABKO, T. E.: phys. stat. sol. (a) **42**, 681 (1977)

REDDY, N. A., KISTAIAH, P., MURTHY, K. S.: High Temperature — High Pressure **15**, 517 (1983)

(Received April 7, 1987)

Authors' addresses:

N. S. ORLOVA
Institute of Physics of Solids and Semiconductors
Academy of Sciences of the BSSR
ul. P. Brovki, 17
220726 Minsk, USSR

I. V. BODNAR
Minsk Radio Engineering Institute
ul. P. Brovki, 6
220069 Minsk, USSR

Cryst. Res. Technol.	22	1987	11	1415–1419

Yu. Berezovskaya, S. Budurov, T. Spasov

Department of Chemistry, University of Sofia, Bulgaria

Specific Heats of FCC Solid Solutions of the Bi_2O_3-Y_2O_3 System

For the first time specific heats of the face centered cubic solid solutions $(Bi_2O_3)_{1-x}(Y_2O_3)_x$ in the temperature range 350–1000 K were measured by means of Differential Scanning Calorimetry. Samples were also investigated with thermal expansion. The results showed the phase transitions for $(Bi_2O_3)_{0.75}(Y_2O_3)_{0.25}$, which is acsribed to $\delta \rightarrow \delta^*$ change in lattice mentioned in literature and for $(Bi_2O_3)_{0.58}(Y_2O_3)_{0.42}$, which is connected with a metastable behaviour of this sample.

1. Introduction

The high-temperature δ-phase of Bi_2O_3 is the best oxide ion conductor known (Takahashi, Iwahara), with a conductivity of $\sim 1\ \Omega^{-1}\ cm^{-1}$ at 1023 K compared to a value of $\sim 10^{-4}\ \Omega^{-1}\ cm^{-1}$ for the widely used ZrO_2–CaO electrolyte (Kilner, Steele). The temperature range over which this phase is stable is, however, rather narrow (Gattow, Schütze) (730–825 °C mp) and electronic conduction predominates below 730 °C where the monoclinic α-phase is stable. Moreover, phase transformation is accompanied by a large volume change which causes the specimen to collapse. Therefore pure δ-Bi_2O_3 is unsuitable for use as a solid electrolyte.

A number of cation dopants lower the temperature of transition from the poorly conducting low-temperature phase to the superionic, δ phase (Takahashi, Iwahara). For example, Y_2O_3-doped Bi_2O_3 is stable with the face-centered-cubic fluorite-like structure at room temperature for dopant levels of between 25 and 43 mol.%. All investigations on this system, carried out until now mainly deal with oxygen ion-conduction and ionic transference number measurements (Takahashi et al. 1975; Changzhen et al.) and with some structural aspects (Datta, Meehan; Battle et al.). Information about thermodynamic data on $(Bi_2O_3)_{1-x}(Y_2O_3)_x$ is very scarce. Takahashi et al. (1977a) determined the standard free energy of formation of $(Bi_2O_3)_{0.73} \cdot (Y_2O_3)_{0.27}$. Lawless et al. (Lawless, Swartz) from the heat capacity and thermo-conductivity measurements in the temperature range 1.7–40 K have reported that $(Bi_2O_3)_{0.75}(Y_2O_3)_{0.25}$ displays classic glass-like thermal properties at low temperatures (below 12 K). No other results concerning thermodynamic properties of $(Bi_2O_3)_{0.25} \cdot (Y_2O_3)_x$ are reported in literature. From these points of view we have investigated temperature and composition dependence of specific heats in sintered oxides of Bi_2O_3–Y_2O_3 system. Observed heat effects were also investigated with thermal expansion. The results were combined with literature data on conductivity.

This investigation is a part of our studies on thermodynamic properties of δ-Bi_2O_3 based solid solutions.

2. Experimental

2.1. Preparation of the specimens

Four samples within stable range of compositions (25—43 mol.% Y_2O_3) were prepared. The preparation method was the same as the technique reported previously (Takahashi et al. 1975). Following Takahashi et al. (1975) the powders of Bi_2O_3 and Y_2O_3 were mixed and calcined at 750—800 °C for 10 h in air, then finely ground and sintered in air for 10 h. The sintering temperature was changed from 880 °C for 25 mol.% Y_2O_3 to 970 °C for 42 mol.% Y_2O_3. X-ray diffraction was carried out using CoKα radiation. The reproducibility of lattice parameter measurement was within ±0.007 Å.

2.2. Specific heats measurement

Specific heats measurements were performed with a Differential Scanning Calorimeter (DSC-2) "Perkin-Elmer" in the range of temperature 350—1000 K.

For highest accuracy in specific heats a specific heat standard run under the same conditions as the unknown samples was used. Synthetic sapphire served as a standard. The ordinate deflection was measured from a "no-sample" baseline.

At any point on the curve, presented on the chart, the ordinate deflection y is equal to $K\dot{T}C_pm$, where

K calibration constant

$\dot{T}$ constant scanning rate

m weight of the sample or standard.

Thus for sapphire $y' = K\dot{T}C'_pm'$ and for the sample $y = K\dot{T}C_pm$. Therefore $C_p = C'_pym'/y'm$.

In general C_p may be obtained to ±2% routinely. The principal source of error is the residual uncertainty in the no-sample baseline. This error can be reduced by scanning over shorter intervals or simply by repetition of the experiment and statistical reduction of the data.

In our case scanning intervals were 100 degrees, scanning rate — 10 deg/min, each run was repeated.

Temperature behaviour of the specific heat for all samples was also investigated by scanning in the whole temperature range (350—1000 K).

2.3. Thermal expansion measurement

The expansion as a function of temperature within 350—1000 K was measured with a Thermomechanical Analyser System (TMS-2) "Perkin-Elmer" (scanning rate: 10 deg per min) on sintered tablets having a diameter of 8 mm and a length of 6 mm.

3. Results and discussion

3.1. Specific heats of the specimens

A survey of the experimental results on specific heats of the solid solutions $(Bi_2O_3)_{1-x}(Y_2O_3)_x$ is given in Figure 1. Numeral values, obtained from the least-squares analysis of the monotonous parts of experimental curves, where no phase or chemical transition occurs, are presented in Table 1. Resulting least squares are of order 10^{-4} for C_p in cal/gr deg. Specific heats as a functions of temperature and composition are shown in Figure 2. In this figure the specific heats of pure Bi_2O_3 (Barin et al.) and Y_2O_3 (Barin, Knacke) are given as a reference material.

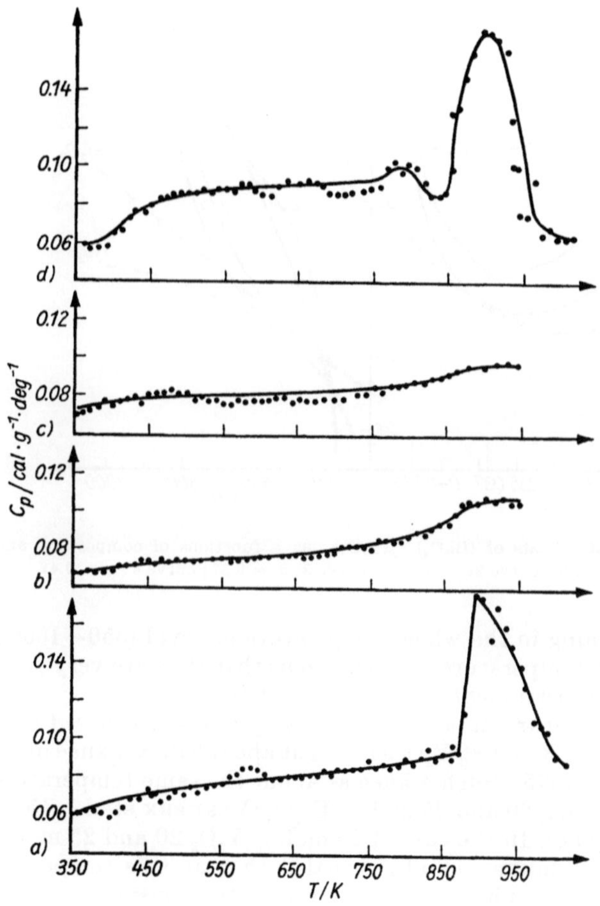

Fig. 1. Specific heats of $(Bi_2O_3)_{1-x}(Y_2O_3)_x$ vs temperature: a) $x = 0.25$; b: $x = 0.30$; c: $x = 0.37$; d: $x = 0.42$

Table 1
Specific heats of the fcc solid solutions $(Bi_2O_3)_{1-x}(Y_2O_3)_x$

x	$C_p = A + B \cdot 10^{-3}T - C \cdot 10^5 \cdot T^{-2} + D \cdot 10^{-6}T^2$ cal/mol K				
	A	B	C	D	temperature range, K
0.25	30.41	6.24	11.69	4.08	350–870
0.30	84.38	−147.56	25.34	113.1	350–950
0.37	97.25	−181.42	27.14	133.31	350–950
0.42	66.81	−50.09	40.05	20.61	350–850

3.2. Phase transitions

As it can be seen from Figure 1, a heat effect was observed at about 870–880 K for $(Bi_2O_3)_{0.75}(Y_2O_3)_{0.25}$. In the thermal expansion measurement for this sample a decrease in the expansion coefficient α from $\sim 1.3 \cdot 10^{-5}$ K^{-1} to $\sim 5.8 \cdot 10^{-6}$ K^{-1} was observed

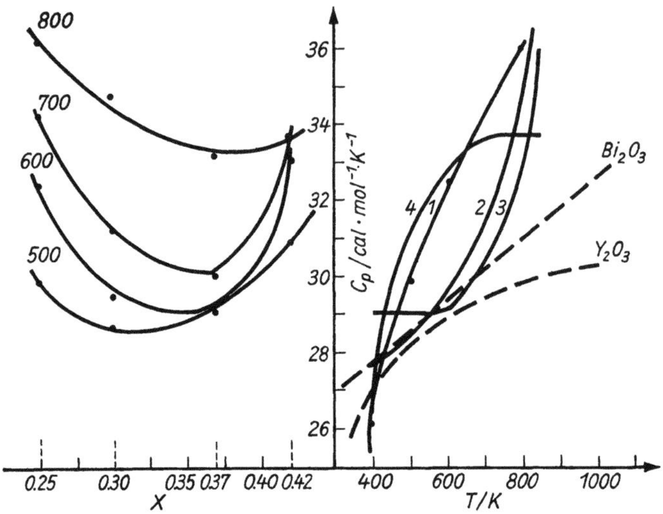

Fig. 2. Specific heats of $(Bi_2O_3)_{1-x}(Y_2O_3)_x$ as a functions of composition and temperature: curve 1: $x = 0.25$; curve 2: $x = 0.30$; curve 3: $x = 0.37$; curve 4: $x = 0.42$

at about 890 K. Scanning in the whole temperature interval (350—1000 K) revealed no heat effects at high temperatures, which means that they are very small compared with the sensitivity of instrument at these temperatures.

According to the literature data on conductivity of δ-Bi_2O_3 based solid solutions, samples containing 25—33 mol.% Y_2O_3 showed at about 870 K a knee in the Arrhenius plot (TAKAHASHI et al. 1975). Such a knee at about the same temperature was found for the samples containing 20 and 25 mol.% Er_2O_3 (VERKERK et al.), 15 mol.% Nb_2O_5 (TAKAHASHI et al. 1977b). In the case of 25 mol.% Y_2O_3 20 and 25 mol.% Er_2O_3 HT Guinier experiments (VERKERK et al.) revealed that at this temperature the fcc-phase δ, transforms to an fcc-phase δ^* with a larger lattice constant; a suggestion about changes in the ordering of oxygen in the lattice was made. Our results confirm this suggestion for $(Bi_2O_3)_{0.75}$ $(Y_2O_3)_{0.25}$.

Specific heats measurements, thermal expansion, scanning showed no heat effects in the whole temperature interval for the samples with 30 and 37 mol.% Y_2O_3.

As it can be seen from Figure 1, a heat effect was observed in the temperature region of about 800—850 K for $(Bi_2O_3)_{0.58}$ $(Y_2O_3)_{0.42}$. In the thermal expansion measurement for this sample a decrease in the expansion coefficient α from $\sim 1.7 \cdot 10^{-5}$ K^{-1} to $\sim 4.9 \cdot 10^{-6}$ K^{-1} was observed at about the same temperatures. Scanning revealed no effect. Obviously, $(Bi_2O_3)_{0.58}(Y_2O_3)_{0.42}$ is metastable and exists at relatively low temperatures as a mixture of two equilibrium phases: fcc solid solution $(Bi_2O_3)_{1-x} \cdot \cdot (Y_2O_3)_x$ with x less than 0.42 and a small amount of Y_2O_3-rich phase.

According to literature data on system Bi_2O_3–Er_2O_3 monophasic fcc structure is stable within 17.5—45.5 mol.% Er_2O_3, but for samples containing 40 mol.% Er_2O_3 an electronic component of conductivity is introduced at low temperatures (VERKERK et al.). The same is observed for Bi_2O_3–Dy_2O_3 system (VERKERK, BURGGRAAF) and from ionic transference number measurement the electrical conductivity has to be ascribed to the second phase.

Consequently, solid solutions based on δ-Bi_2O_3 with a large percent of dopant oxide may be metastable.

4. Conclusions

Specific heats of solid solutions of the Bi_2O_3–Y_2O_3 system are measured. For $(Bi_2O_3)_{0.75}(Y_2O_3)_{0.25}$ a phase transition is observed at about 870–890 K, confirmed by thermal expansion measurement. This transition can be ascribed to the $\delta - \delta^*$ change, mentioned in literature. For $(Bi_2O_3)_{0.58}(Y_2O_3)_{0.42}$ a phase transition of first order is observed at about 800–850 K, also confirmed by thermal expansion measurement. We conclude that this sample is metastable and exists at low temperatures as a mixture of two phases.

References

BARIN, I., KNACKE, O.: Thermochemical properties of inorganic substances, Springer-Verlag, Berlin, Heidelberg, New York 1973
BARIN, I., KNACKE, O., KUBASCHEWSKI, O.: Thermochemical properties of inorganic substances, Springer-Verlag, Berlin, Heidelberg, New York 1977
BATTLE, P. D., CATLOW, C. R. A., HEAP, J. W., MORONEY, L. M.: J. Solid State Chem. **63**, 8 (1986)
CHANGZHEN, W., XIUGUANG, X., BAOZHEN, Z.: Solid State Ionics **13**, 135 (1984)
DATTA, R. K., MEEHAN, J. P.: Z. anorg. allg. Chem. **383**, 328 (1971)
GATTOW, G., SCHÜTZE, D.: Z. anorg. allgem. Chem. **318**, 176 (1962)
KILNER, J. A., STEELE, B. C. H.: Non stoichiometric oxides (O. T. SORENSEN, Ed.), Chap. 3, New York 1981
LAWLESS, W. N., SWARTZ, S. L.: Phys. Rev. B. **28**, 2125 (1983)
TAKAHASHI, T., ESAKA, T., IWAHARA, H.: J. Appl. Electrochem. **7**, 299 (1977) a
TAKAHASHI, T., ESAKA, T., IWAHARA, H.: J. Electrochem. Soc. **124**, 1563 (1977) b
TAKAHASHI, T., IWAHARA, H.: Mat. Res. Bull. **13**, 1447 (1987)
TAKAHASHI, T., IWAHARA, H., ARAO, T.: J. Appl. Electrochem. **5**, 178 (1975)
VERKERK, M. J., BURGGRAAF, A. J.: J. Electrochem. Soc. **128**, 75 (1981)
VERKERK, M. J., KEIZER, K., BURGGRAAF, A. J.: J. Appl. Electrochem. **10**, 81 (1980)

(Received April 14, 1987)

Authors' address:

Yu. BEREZOVSKAYA, S. BUDUROV, T. SPASOV
Department of Chemistry, University of Sofia
1 Anton Ivanov Str.
1126 Sofia, Bulgaria

Book Review

JOHN A. DEAN

Handbook of Organic Chemistry

McGraw-Hill Book Company New York u. a., 1987, 976 Seiten, 22 Abbildungen, 141 Tabellen, ISBN 0-07-016193-3, DM 177.40

Das vorliegende Handbuch von JOHN A. DEAN ist ein Tabellenbuch speziell für den organisch arbeitenden Chemiker. Der Autor hat seine jahrelangen Erfahrungen als Herausgeber des „LANGE's Handbook of Chemistry" hervorragend genutzt, um ein einbändiges Kompendium zusammenzustellen, daß für den Organiker mehr bietet, als vergleichbare Tabellenbücher. Das betrifft nicht die Anzahl der in der üblichen Form (Name, Strukturformel, MG, Beilsteinzitat, n_D^{20}, Fp., Kp., Flammp., Löslichkeit) beschriebenen organischen Verbindungen, mit ca. 4000 verschiedenen organischen und knapp 1000 anorganischen Substanzen liegt es sogar hinter normalen Tabellenbüchern zurück. Aber die auf den anderen Seiten aufgenommenen Tabellen sparen dem Benutzer weitere bisher notwendige Sammelwerke und Übersichten. Besonders erwähnt werden sollen an dieser Stelle: 1. Die spektroskopischen Daten, die auf 114 Seiten aufgeführt werden. Neben ausführlichen Angaben zur UV/VIS-, IR- und Raman-Spektroskopie finden sich umfangreiche Tabellen zu ^{1}H-, ^{13}C-, ^{11}B-, ^{15}N-, ^{19}F-, ^{29}Si- und ^{31}P-chemischen Verschiebungen einschließlich Kopplungskonstanten. 2. Von ca. 200 Substituenten werden die Hammett- und Taft-Substituentenkonstanten angegeben. 3. Im Abschnitt 3 werden zahlreiche besonders für den Organiker wichtige Bindungslängen und -energien aufgeführt.

Im Einzelnen ist das Handbuch in 11 Abschnitte gegliedert: 1. Organische Verbindungen 384 S., 14 Tab., 2. Anorganische Verbindungen 33 S., 1 Tab., 3. Eigenschaften von Atomen, Atomgruppen und Bindungen 30 S., 7 Tab., 4. Physikalische Eigenschaften 88 S., 13 Tab., 5. Thermodynamische Eigenschaften 92 S., 4 Tab., 6. Spektroskopie 114 S., 67 Tab., 7. Lineare Freie Energie Beziehungen 10 S., 4 Tab., 8. Elektrolyte, Chemisches Gleichgewicht 88 S., 11 Tab., 9. Laboratoriumsmethoden 11 S., 6 Tab., 10. Polymere, Gummi, Fette, Öle, Wachse 73 S., 7 Tab., 11. Verschiedenes (Physikal. Konstanten, Umrechnungsfaktoren u. a.) 15 S., 7 Tab.

Vervollständigt wird das Handbuch durch eine recht ausführliche Einführung in die Nomenklatur organischer Verbindungen mit stereochemischen Grundlagen ausgehend von den IUPAC-Regeln, sowie zahlreiche kurze, prägnante theoretische Erläuterungen zu einzelnen Abschnitten. Der hervorragende Druck und die sehr übersichtliche Anordnung der Tabellen machen das Auffinden bestimmter Daten problemlos. Auch die getrennte Seitenzählung nach Abschnitten ist außerordentlich nutzerfreundlich. Schon nach kurzer Benutzungszeit ist das ohnehin sehr knapp gehaltene Sachverzeichnis nicht mehr notwendig.

Der „Kleine Dean" ist ein Handbuch, das in erster Linie für den täglichen Gebrauch im organischen Laboratorium gedacht ist, dafür ist er ausgezeichnet geeignet. Man kann ihm nur noch viele Auflagen wünschen. K. SCHULZE

Cryst. Res. Technol.	22	1987	11	1421–1424

Yu. Berezovskaya, S. Budurov, T. Spasov

Department of Chemistry, University of Sofia, Bulgaria

Specific Heats of FCC Solid Solutions of the Bi_2O_3–Er_2O_3 System

For the first time specific heats of the fcc solid solutions $(Bi_2O_3)_{1-x}(Er_2O_3)_x$ in the temperature range 350–1000 K were measured by means of Differental Scanning Calorimetry. Samples were also investigated with thermal expansion. A possibility of tetragonal distortion with subsequent transformation in δ^* phase is supposed for $(Bi_2O_3)_{0.80}(Er_2O_3)_{0.20}$. Thermodynamic data on $(Bi_2O_3)_{1-x}(Er_2O_3)_x$ compared with ideal solutions are discussed.

1. Introduction

In recent years much work has been directed to the development of materials which predominantly exhibit oxygen ion conduction at low temperatures. Of these materials, several Bi_2O_3-based face centered cubic oxides are reported to be amongst the best oxide ion conductors known (Takahashi et al. 1975a, b; Verkerk, Burggraaf 1981). System Bi_2O_3–Er_2O_3, where fcc structure is stable for samples containing 17.5 to 45.5 mol.% Er_2O_3 is one of the most interesting. The lower boundary of this composition range is the smallest percentage valence (111) substituent known in literature which can stabilize the fcc structure at low temperatures (Verkerk et al.). The conductivity values for $(Bi_2O_3)_{0.80}(Er_2O_3)_{0.20}$ are 2–3 times higher than the best oxygen ion conductor reported for substituted Bi_2O_3 systems and 50–100 times higher than those of stabilized zirconia. In literature the results of oxygen ion conduction and ionic transference number measurements on fcc $(Bi_2O_3)_{1-x}(Er_2O_3)_x$ solid solutions are reported (Verkerk et al.). Results concerning the phase diagram and occurrence of the δ-phase at low temperatures (Datta, Meehan; Nasonova et al.) are rather contradictory. Information about thermodynamic data on $(Bi_2O_3)_{1-x}(Er_2O_3)_x$ is practically absent. Only standard free energy of formation of $(Bi_2O_3)_{0.545}(Er_2O_3)_{0.455}$ in the temperature range 750–950 K, obtained by electrochemical measurement, is determined (Verkerk, Burggraaf 1980). From these points of view we have investigated temperature and composition dependence of specific heats in sintered oxides of the Bi_2O_3–Er_2O_3 system. Thermal expansion measurements also have been carried out.

Present work is a part of our studies on thermodynamic properties of δ-based Bi_2O_3 solid electrolytes.

2. Experimental

2.1. Preparation of the specimens

Four samples within stable range of compositions (17.5–45.5 mol.% Er_2O_3) were prepared. The preparation method was the same as the technique reported previously (Vorkerk et al.). Following Verkerk et al. the powders of Bi_2O_3 and Er_2O_3 were thoroughly mixed and pre-fired at 1020—1120 K for 16 h, finely ground and sintered in air for 65 h and cooled down to room temperature at 0.5 K min^{-1}. X-ray diffraction was carried out using CoKα radiation; analysis showed samples to consist of a fcc single phase. The reproducibility of lattice parameter measurement was within ± 0.007 Å.

2.2. Specific heats measurement

Specific heats measurements were performed with a Differential Scanning Calorimeter (DSC-2) "Perkin-Elmer" in the range of temperature 350—1000 K.

For highest accuracy in specific heats a specific heat standard run under the same conditions as the unknown sample was used. Synthetic sapphire served as a standard. The ordinate deflection was measured from a "no-sample" baseline.

At any point on the curve, presented on the chart, the ordinate deflection y is equal to

$K\dot{T}C_pm$,

where

K calibration constant

$\dot{T}$ constant scanning rate

m weight of the sample or standard.

Thus for sapphire $y' = K\dot{T}C_p'm'$ and for the sample $y = K\dot{T}C_pm$. Therefore $C_p = C_p'ym'/my'$.

In general C_p may be obtained to $\pm 2\%$ routinely. The principal source of error is the residual uncertainty in the no-sample baseline. This error can be reduced by scanning over shorter intervals or simply by repetition of the experiment and statistical reduction of the data. In our case scanning rate was 10 deg/min, scanning intervals were 100 degrees, each run was repeated.

Scanning in the whole temperature range (350—1000 K) by means of special Thermal Analyses Programme has also been carried out.

2.3. Thermal expansion measurement

The expansion as a function of temperature in the range 350—1000 K was measured with a Thermomechanical Analyser System (TMS-2) "Perkin-Elmer" (scanning rate: 10 deg/min) or sintered tablets.

3. Results and discussions

A survey of the experimental results on specific heats of the investigated solid solutions $(Bi_2O_3)_{1-x}(Er_2O_3)_x$ is given in Figure 1. Numeral values obtained from the least-squares analyses of the curves are presented in Table 1. Resulting least squares are of order 10^{-4} for C_p in cal/gr deg. Specific heats as a functions of temperature and composition are shown in Figure 2. In this figure the specific heats of pure Bi_2O_3 and Er_2O_3 (both Barin et al.) are given as a reference material.

For the samples containing 27, 35, and 42 mol.% Er_2O_3 specific heats measurements, scanning and thermal expansion revealed no heat effects in the whole temperature range investigated, which confirms the existence of a single fcc phase within these intervals of composition and temperature. For $(Bi_2O_3)_{0.80}(Er_2O_3)_{0.20}$ specific heat measurement and scanning also showed no phase transitions within 350—1000 K, but the coefficient of thermal expansion becomes temperature dependent at about 790—880 K. According to the literature data (Verkerk et al.) in the case of samples containing 15 and 17.5 mol.% Er_2O_3 a tetragonal distortion ($\delta \rightarrow \beta^*$) starts at 730 K, which could not be measured with HT Guinier and thermal expansion. At 840 K the value of tetragonal distortion is large enough to be measured and increase strongly

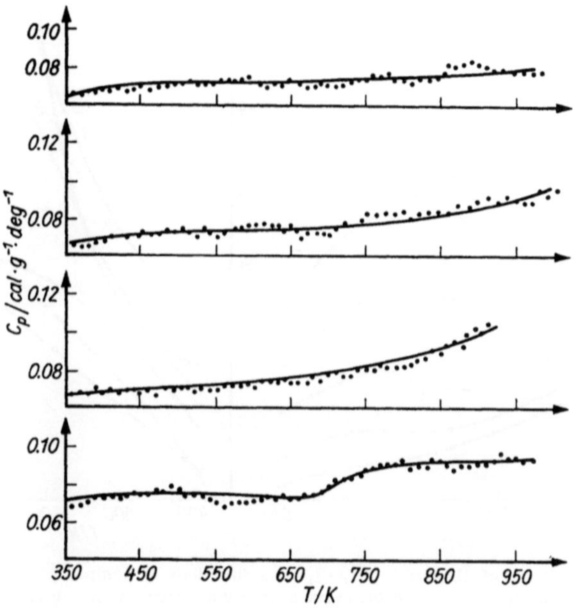

Fig. 1. Specific heats of $(Bi_2O_3)_{1-x}(Er_2O_3)_x$ vs temperature: a: $x = 0.20$; b: $x = 0.27$; c: $x = 0.35$; d: $x = 0.42$

Table 1
Specific heats of the fcc solid solutions $(Bi_2O_3)_{1-x}(Er_2O_3)_x$

x	$C_p = A + B \cdot 10^{-3}T + C \cdot 10^5 T^{-2} + D \cdot 10^{-6}T^2$ cal/mol K				
	A	B	C	D	temperature range, K
0.20	19.14	15.96	8.16	8.68	350–1000
0.27	82.48	−147.0	−19.39	119.2	350–950
0.35	12.42	33.13	8.50	−6.01	350–1000
0.42	0.42	66.35	12.78	−35.02	350–1000

with temperature, which agrees with the thermal expansion data. At about 880 K the β^*-phase transforms into an fcc phase (δ^*), which has a larger lattice constant than the δ-phase. But DTA revealed no heat effects, which means that they are either very small or are smeared out.

Perhaps, such tetragonal distortion with subsequent transformation in δ^* phase takes place in the case of $(Bi_2O_3)_{0.58}(Er_2O_3)_{0.20}$, although it is not confirmed by scanning and specific heat measurements.

VERKERK and BURGGRAAF (1980) resulting from electrochemical measurement and taking into account the standard deviation in the experimental values find a good agrreement between the thermodynamic values of $(Bi_2O_3)_{0.545}(Er_2O_3)_{0.455}$ and pure δ-Bi_2O_3 and suggest "ideal" behaviour of stabilized bismuth sesquioxide.

From our results $\Delta C_p = C_p - (1 - x)\, C_p^0(Bi_2O_3) - xC_p^0(Er_2O_3)$ for $(Bi_2O_3)_{0.58} \cdot (Er_2O_3)_{0.42}$ is near one in cal/mol. K in the whole temperature range. Taking into account the standard deviation, solid solution with 42 mol.% Er_2O_3 does not differ in large extent from "ideal" one. But this suggestion cannot be applied to the other

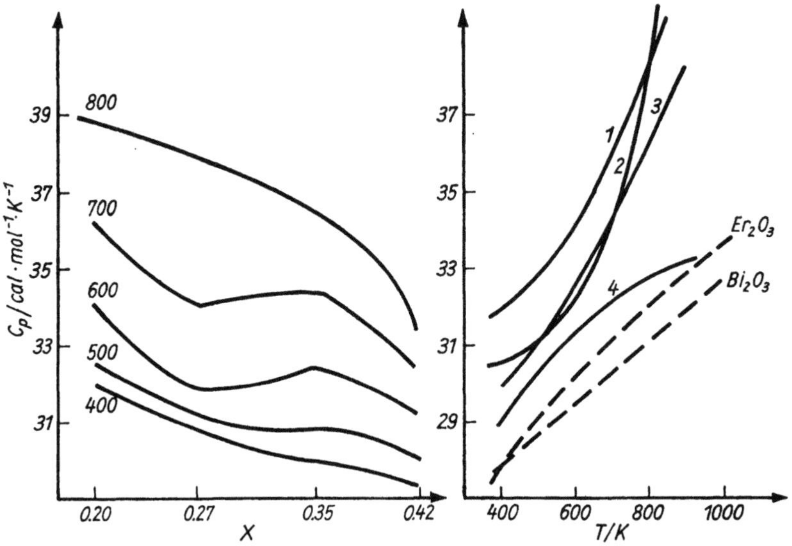

Fig. 2. Specific heats of $(Bi_2O_3)_{1-x}(Er_2O_3)_x$ as a functions of composition and temperature: curve 1: $x = 0.25$; curve 2: $x = 0.30$; curve 3: $x = 0.37$; curve 4: $x = 0.42$

solutions of the Bi_2O_3–Er_2O_3 system, where values of ΔC_p are small but are strongly temperature dependent.

4. Conclusions

Specific heats of solid solutions of the Bi_2O_3–Er_2O_3 system are measured. Resulting from thermal expansion a possibility of tetragonal distortion with subsequent transformation in δ^* phase, mentioned in literature for the samples with 15 and 17.5 mol.% Er_2O_3 is supposed for $(Bi_2O_3)_{0.80}(Er_2O_3)_{0.20}$ in the temperature range 790–880 K. Thermodynamic behaviour of investigated solid solutions compared with ideal solutions is discussed.

References

BARIN, I., KNACKE, O., KUBASCHEWSKI, O.: Thermochemical properties of inorganic substances, Supplement, Berlin, Heidelberg, New York 1977

DATTA, R. K., MEEHAN, J. P.: Z. anorg. allg. Chem. **383**, 328 (1971)

NASONOVA, S. N., SEREBRENNIKOV, V. V., NARNOV, G. A.: Russ. J. Inorg. Chem. **18**, 1244 (1973)

VERKERK, M. J., BURGGRAAF, A. J.: J. Appl. Electrochem. **10**, 677 (1980)

VERKERK, M. J., BURGGRAAF, A. J.: J. Electrochem. Soc. **128**, 75 (1981)

VERKERK, M. J., KEIZER, K., BURGGRAAF, A. J.: J. Appl. Electrochem. **10**, 81 (1980)

TAKAHASHI, T., ESAKA, T., IWAHARA, H.: J. Appl. Electrochem. **5**, 197 (1975) b

TAKAHASHI, T., IWAHARA, H., ARAO, T.: J. Appl. Electrochem. **5**, 187 (1975) a

(Received April 14, 1987)

Authors' address:

YU. BEREZOVSKAYA, S. BUDUROV, T. SPASOV
Department of Chemistry, University of Sofia
Anton Ivanov Str. 1
1126 Sofia, Bulgaria

Cryst. Res. Technol. | 22 | 1987 | 11 | 1425–1429

T. Chassé, W. Heichler, J. Langhammer, W. Zwanzig

Martin-Luther-Universität Halle-Wittenberg, Sektion Physik

Sputter Depth Profiling of GaAlAs Double Heterostructures Using Auger Electron Spectroscopy

Sputter depth profiles of GaAlAs double heterostructures had been taken by the help of Auger electron spectroscopy (AES). The sputtering ion beam consisted of either N^+ — or Ar^+ ions. In the case of using N^+ ions an essentially better depth resolution was found comparing with Ar^+ ions, especially in a sputter depth of some μm. Investigations of the sputter crater using Scanning Electron Microscopy (SEM) and Electron Probe Micro Analysis (EPMA) showed, that the improved depth resolution during N_2^+ ion sputtering is a result of the essential lowered sputter induced surface roughness. This should be caused by the stronger amorphisation of the target surface using N_2^+ ions which is a consequence of the chemical reactions taking place at the surface with the sputtering ions.

Introduction

A comparison of depth resolution values (Δz) in profiling "sandwich structures" from various semiconducting materials and metals given by Hofmann (1981) shows that sputtering using noble gas ions gives in most cases better values for semiconductor structures than for metal "sandwich structures". Blattner et al. found better values for Δz while sputtering metal systems with N_2^+ ions compared to Ar^+-sputtering. The same effect was established by Heichler et al. at Cu–Ni interfaces at a sputter depth of some 100 Å.

These results are promising to expect a further improvement of Δz by sputtering with N_2^+ ions also in the case of semiconductor "sandwich-structures" especially at large sputter depths of some μm. That is important for the depth profiling of the Al concentration gradient in GaAlAs double heterostructures, because the interesting region is at a sputter depth of 2—3 μm.

Experimental

The experiments for compositional depth profiling were carried out in a conventional stainless-steel self-made UHV equipment (base pressure $\sim 10^{-7}$ Pa) with a one stage cylindrical mirror analyzer (energy resolution $\Delta E/E = 0.5\%$, Brümmer et al.; Heichler et al.). The Auger electrons were excited by an electron beam with the parameters 60 μm, $I = 2$ μA, angle of incidence to the target surface 20°. The ion beam is generated by a Penning-ion source. In order to minimize the development of sputter induced surface roughness, the angle of incidence of the ion beam (measured to the surface) was lowered to 30°. The surface concentration of the chemical components were calculated using the Palmberg model, which proposes homogeneous distribution (Davis et al.) in the range of the mean escape depth of the Auger electrons, according to the following equation

$$C_x = \frac{H_x/S_x}{\sum_i H_i/S_i},$$

with S_i sensitivity factor and H_i peak-to-peak heights of the corresponding Auger lines.

The sensitivity factors S_{Ga} and S_{As} of GaAs had been derived from the (110) surface cleaved in UHV. They are the same as those given by DAVIS et al. That's why we used the sensitivity factor for Al from DAVIS et al. without further experimental checking. Because of the high secondary electron background in the low energy region $E < 70$ eV (low angle of incidence of primary electron beam) only the high energy lines were used. The peak-to-peak heights of the Ga-1070 eV, the As-1224 eV and the Al-1380 eV lines had been derived from the spectra recorded in the derivative mode without further correction.

Results and discussion

The layer structure of the investigated LPE-grown specimens is shown in Figure 1. The first specimen called type A was cut in two samples. An optical investigation showed no surface defects and a minimum density of growth terraces. The compositional depth profiles were measured under identical conditions. Figures 2a and 2b show the corresponding profiles using Ar^+- or N_2^+ ions. Δz_1 has been estimated as the

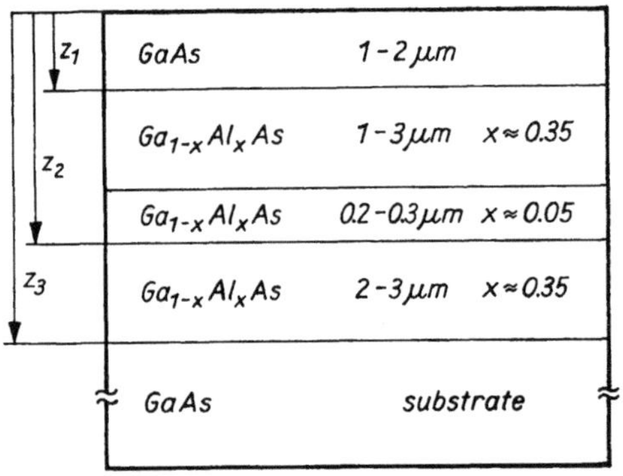

Fig. 1. Layer-structure of the investigated double heterostructure

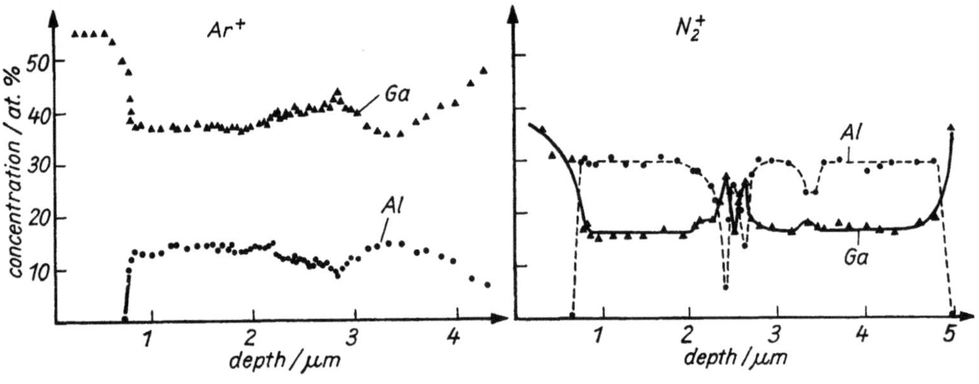

Fig. 2. a) Al- and Ga concentration depth profile obtained for specimen A using Ar^+ sputter ions; b) Al- and Ga concentration depth profile obtained for specimen A using N^+ sputter ions

depth range, which corresponds to a change from 84 to 16% of the maximum Al concentration. The profile from Figure 2a gives a value $\Delta z_1 = 56$ nm at the interface contact layer — first GaAlAs layer.

The depth resolution in the case of Ar^+ sputtering decreases drastically with increasing depth. The active layer is only slightly indicated and the interface second GaAlAs layer-substrate is almost not locatable.

Investigation of the sputter crater after sputter profiling by SEM showed two types of surface roughness: a few of "mesa-like" structures, (Fig. 4) and a lot of so called "cones" and "pits", (Fig. 5). The slope angle of the side walls of the "cones" and "pits" corresponds to the incidence angle of the Ar^+ ions. This is not the case for the "mesa-like" structures. The height of the "mesa-like" structures is nearly the thickness of the whole layer structure, e.g. some μm. That's why one has to expect that they are crossing all layers. Using EPMA (∅ 1 μm) there was not found any Al on the top and at the side walls of the "mesa-like" structures. Therefore these structures are probably growth defects which accidentally arise during the layer growth. A possible mechanism is known as "melt carry over" (MAHAJAN). On the top of the microstructures (Fig. 5) Al was detected, but not around them. This means that the bottom of the sputtering crater is within the substrate material and the top of the "cones" and "pits" lies in the last GaAlAs layer. Caused by the existence of these structures

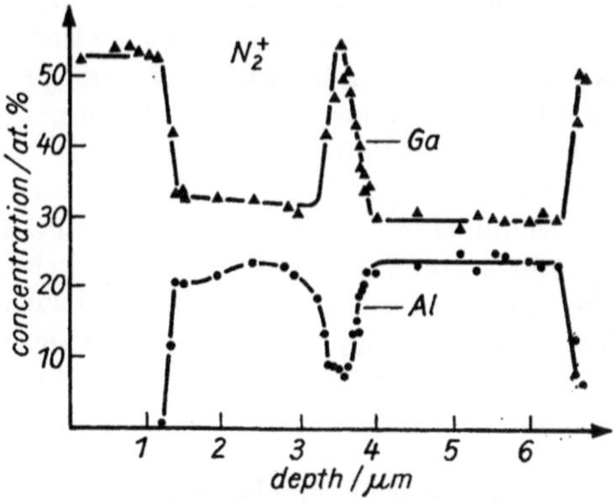

Fig. 3. Al- and Ga concentration depth profile obtained for specimen B using N_2^+ sputter ions

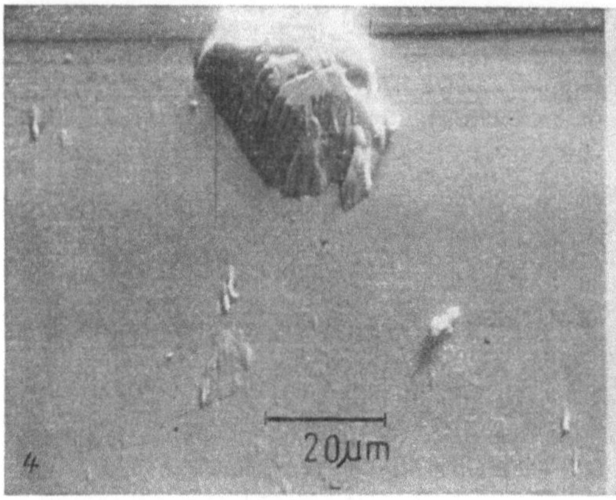

Fig. 4. "mesalike" structures on the sputter crater ground after Ar^+ sputtering (specimen A)

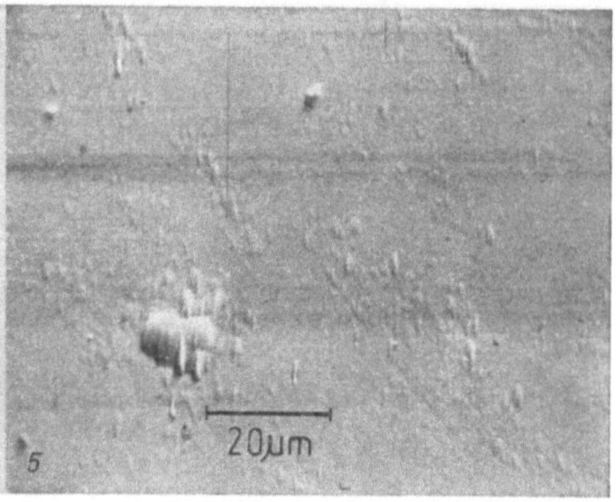

Fig. 5. microstructures on the sputter crater ground after Ar^+ sputtering (specimen A)

the interface GaAlAs layer-substrate is smeared out during depth profiling, because the diameter of the exciting AES electron beam is 60 μm and the characteristic diameter of the "cones" and "pits" is much smaller, e.g. some μm.

The depth resolution during N_2^+ ion sputtering deteriorates not so strongly with increasing sputter depth as in the case of Ar^+ ion sputtering (Fig. 2b). No corresponding structures like these in Figures 4 and 5 had been found with SEM at the bottom of the N_2^+ sputtering crater. A qualitatively very similar Al concentration depth profile of the active layer was found using SIMS (TRAPP, RICHTER) which showed the same Al-concentration dip in the layer with $x = 0.05$ Al.

The specimen B was in the same manner sputter-profiled as specimen A but with a higher time resolution at the interface. The following results were obtained for Δz:

contact layer — first GaAlAs layer	$\Delta z_1 = 67.5$ nm	$z_1 = 1.25$ μm
active layer — second GaAlAs	$\Delta z_2 = 120$ nm	$z_2 = 3.62$ μm
second GaAlAs layer-substrate	$\Delta z_3 = 143$ nm	$z_3 = 6.52$ μm .

Neglecting the sputter rate dependence of the actual chemical composition, we found that a dependence $\Delta z \sim \sqrt{z}$ fits experimental data for Δz satisfactorily. Such a type of dependence can be expected, if Δz is determined mainly by sputter induced roughening of the crater surface (HOFMANN 1982). This indicates that also during N_2^+ sputtering roughening of the sample surface takes place, which is comparatively small, with respect to Ar^+ sputtering.

Due to nonuniform ion current density, the bottom of the final sputter crater (depth = 6.52 μm) had an angle with respect to the surface, which was measured by the inclination of reflected laser beam. The average slope angle was 0.4°. Hence the sputter crater (∅ 4 mm) intersected the layer structure with a very small angle and the whole structure may be observed with an optical microscope after sputtering down. This allowed us knowing the layer thickness to measure the angle independently and the first value was established. In the consequence the nonuniform sputter current density caused a contribution to the measured value Δz, due to the inclined intersection of the layer structure. The value of this contribution depends on the ratio of the diameter of the electron and ion beams and the electron beam position relative to the crater centre (DUNCAN et al.). So the measured values Δz for the different interfaces are not the lowest possible ones, which could be optimally obtained for N_2^+ sputtering.

The suppression of sputter induced surface roughness for N_2^+ sputtering had been observed also for O_2^+ sputtering. The given explanation there, that this is caused by a higher amorphisation level of the sputter-altered surface layer, due to the reactive nature of the O_2^+ ions holds also in the case of N_2^+ ions. It is well known, that chemical bonds between Ga–N and Al–N are strong so that the precondition for amorphisation is fulfilled. The main effect of amorphisation is to prevent the build-up of sputter induced surface roughening and the initiation of this process by bulk defects. Due to the chemical reactivity of the N_2^+ ions, the selectivity of the sputtering process is altered strongly for comparison to Ar^+ sputtering. In the consequence the quantitative chemical analysis is complicated in the sense, that the surface chemical composition does not reflect the bulk composition in direct manner. On the other hand the increased Al surface concentration, caused by the increased selectivity of N_2^+ sputtering, is advantageously for the depth profiling of the Al concentration gradient in a double-heterostructure (Al bulk concentration 16 at.%, Al-surface concentration during N_2^+ sputtering 25—30 at.%).

Conclusions

The application of N_2^+ ions in the AES sputter depth profiling of GaAlAs double-heterostructures yields an essentially improved depth resolution in a sputter depth up to 6.5 μm compared to Ar^+ depth profiling. Investigations of the Ar^+ sputtered sample surface on the crater bottom in a depth of 6.5 μm by SEM and EPMA showed that this fact is caused by an extreme sputter induced surface roughening. The experimentally obtained dependence $\Delta z \sim \sqrt{z}$ for N_2^+ sputtering shows that there also exists a sputter induced surface roughening, which is small compared to Ar^+ sputtering. The increase of Δz with depth is sufficient small, so that Al concentration profiles in GaAlAs double-heterostructures can be taken with a much better depth resolution extended to large depths. The main reason for the lowering of the sputter induced surface roughening using N_2^+ ions is thought to be the higher amorphisation level.

References

Blattner, R. J., Nadel, S., Evans, C. A., Braundmeier, A. J., Magee, Ch. W.: Surf. Interface Anal. 1, 32–35 (1979)

Brümmer, O., Heichler, W., Langhammer, J., Wippermann, R., Zwanzig, W.: Beiträge zur 6. Tagung Mikrosonde, Dresden 1984

Davis, L. E., MacDonald, N. C., Palmberg, P. W., Riach, G. E., Weber, R. E.: Handbook of Auger Electron Spectroscopy, Physical Electronics Industries 1972

Duncan, S., Smith, R., Sykes, D. E., Walls, J. M.: Surf. Interface Anal. 5, 71–76 (1983)

Heichler, W., Langhammer, J., Wippermann, R., Zwanzig, W.: Wiss. Z. Univ. Halle, XXX′ 82M, H. 4, 71

Hofman, S.: Springer Ser. Chem. Phys. 18 (1982) SIMS III, p. 186

Hofman, S.: Analysis 9, 181–87 (1981)

Mahajan, S.: Micros. Semicond. Mat. Conf., ed. Cullis, A. G. et al. Oxford (1983), p. 259 to 272

Trapp, M., Richter, C.-E.: private communication

(Received April 20, 1987)

Authors' addresses:

Dr. T. Chassé
Sektion Chemie der Karl-Marx-Universität Leipzig,
Analytisches Zentrum
Linnéstr. 2
DDR-Leipzig 7010

Address for correspondence:

Dr. W. Heichler, Dipl.-Phys. J. Langhammer, Prof. Dr. sc. W. Zwanzig
Sektion Physik der Martin-Luther-Universität Halle-Wittenberg
Friedemann-Bach-Platz 6, Postfach
Halle
DDR-4020

Book Review

A. HAHN, W. BOSUM

Geomagnetics — Selected Examples and Case Histories

Geoexploration Monographs, Serie 1, no. 10, S. SAXOV, H. C. SOFFEL (eds)

Gebrüder Bornträger, Berlin, Stuttgart 1986, 91 Fig. und 8 Tab., DM 98.00,
ISBN 3-443-13016X

Die vorliegende Monographie über Geomagnetik ist kein Lehrbuch dieser Disziplin. Sie ist vielmehr ein Bericht über Fortschritte an ausgewählten Beispielen, über Anwendungsmöglichkeiten und Grenzen dieser Methode der angewandten Geophysik nach dem gegenwärtigen Stand. Dieses Konzept ist sehr zu begrüßen.

Auf 166 Seiten werden folgende Probleme und Ergebnisse in acht größeren Abschnitten behandelt:

1) Einführung: Zur Interpretation geomagnetischer Anomalien im Hinblick auf quantitative Angaben der magnetischen Störkörper im Untergrund (Lage, Gestalt, Magnetisierung und strukturelle Situation). Der Zukunft bleibt es überlassen, detaillierte geologisch-petrographische Parameter zu ermitteln, wie z. B. das Alter, die Metamorphose oder Genese der Gesteine im Störkörper. Vorerst ist das noch nicht (oder nur in Ausnahmefällen) möglich.

2) Anomalien über Magnetiterzen, Erzprospektion: Aeromagnetische Messungen über Magnetiterz-Vorkommen bei Kiruna (Schweden). Bodenanomalien über der Eisenerzlagerstätte bei Tschogart (Iran).

3) Anomalien magmatischer Gesteine. Eine besondere Rolle spielen basaltische Gesteine als Decken, Gänge und Stöcke (Pipes). Beispiele werden u. a. gegeben von Minas Geraes (Brasilien), aus dem Saar-Nahe-Trog (Intrusionen im Rotliegenden) und weitere. Hier wird auch über paläomagnetische Studien berichtet.

4) Eine größere Rolle spielen Anomalien über dem Basement, dem kristallinen Untergrund. Es werden zahlreiche Beispiele wiedergegeben, die typisch sind für das Kristallin in flacher wie in sehr tiefer Lage.

5) Magnetische Anomalien über sedimentären Gesteinkomplexen treten — sachlich bedingt — stärker zurück. Dennoch wird darauf verwiesen, daß mit hoch empfindlichen Magnetometern durchaus die Chance besteht, in Sedimentbecken Strukturen zu ermitteln, die auch Kohlenwasserstoffe führen können.

6) Magnetische Anomalien in Gebieten mit geothermischen Besonderheiten. Diskutiert wird die magnetische Anomalie von Soest als Paläo-Phänomen, ferner als rezente geothermische Anomalie die von Olkaria (Kenya).

7) Untersuchungen der Struktur der Erdkruste im Bereich eines alten Schildes, nämlich dem Ukrainischen Schild. Die Ergebnisse werden in einer regionalen magnetischen Karte des Ukrainischen Schildes wiedergegeben. Die Interpretation zieht die seismischen Resultate heran, es werden aber auch die magnetischen Eigenschaften und Modelle betrachtet. Abschließend wird die tektonische Struktur des Ukrainischen Schildes mit dessen magnetischen Anomalienfeld vergleichend interpretiert.

8) Großmaßstäbliche geomagnetische Anomalien nach Satelliten-Beobachtungen.

Dieses letzte Kapitel vorliegenden Buches ist besonders interessant, weil es sowohl auf die verschiedenen Unternehmungen wie deren Resultate eingeht. Man erhält dank verschiedener Interpretationsmodelle einen Überblick über das globale geomagnetische Feld und nicht zuletzt auch über die Krustenmodelle.

Den Autoren wie dem Verlag ist für dieses Buch zu danken, das jedem interessierten Leser, vor allem den Geowissenschaftlern nur empfohlen werden kann. Macht es doch in einem durch die Seismik geprägten Zeitabschnitt auf Wert und Fortschritte der Geomagnetik aufmerksam.

R. LAUTERBACH

Cryst. Res. Technol.	22	1987	11	1431–1435

L. DRESSLER, U. GRIEBNER, R. KITTNER

Friedrich-Schiller-Universität Jena, Sektion Physik

Precision Measurement of Lattice Parameters in LiF Monocrystals

The methodic possibilities of measurements by the Bond-method for a highly precise determination of the lattice parameters in block structures with small block desorientation angles in the region of angular seconds or angular minutes (LiF crystals) have been examined. The lattice distorsion in two planes perpendicular to the (001) surface of crystalis described. The same planes have also been examined by elastooptic methods. In both planes we got for the lattice parameters: a_1 & $a_2 > a_3$ (a_1 and a_2 parallel to the surface of crystal, a_3 perpendicular to the surface of crystal); i.e. we could observe a dilatation of the unit cell parallel to the surface of the crystal. Special results concerning the distribution of local lattice parameters along the surface of sample are discussed in connection with the results of elastooptic observations.

Es werden die meßmethodischen Möglichkeiten des Bond-Verfahrens zur hochpräzisen Gitterparameterbestimmung an Blockstrukturen mit geringen Blockverschwenkungen im Bereich von Winkelsekunden oder -minuten (LiF-Kristalle) untersucht. Dabei wird die Gitterverzerrung in zwei Meßebenen (senkrecht zu einer (001)-Oberfläche) beschrieben, wobei die Meßebenen senkrecht zueinander stehen und dieselben Meßebenen auch polarisationsoptisch untersucht werden.
Die Gitterparametermessung ergab a_1 & $a_2 > a_3$ in beiden Meßebenen (a_3 senkrecht zur Kristalloberfläche), d. h. es tritt eine Längsverformung der Elementarzelle parallel zur Kristalloberfläche auf. Spezielle Ergebnisse der ortsabhängig gemessenen Gitterparameter werden unter Berücksichtigung der polarisationsoptischen Beobachtungen diskutiert.

1. Precise measurement of lattice parameters by the extended BOND-method

The BOND-method is an X-ray diffraction technique for measuring geometric lattice parameters with high accuracy. The principle of the BOND-method (Fig. 1) consists in a highly precise angular measurement between two different positions of crystal (R1 and R2) for which the peak positions of reflections are determined at one and the same atomic plane at R1 and after a rotation at R2. From the peak positions of both counting-rate curves (at R1 and R2) the Bragg angle θ_B results. Supposing a precise knowledge of the value of the X-ray wavelength λ it is possible to calculate the distance d_{hkl} of atomic planes and the length of the unit cell for a known crystal structure. The disadvantage of the BOND-method is that it allows to determine the amount of only one lattice vector at a position of the crystal during one measuring process.

On the basis of this method a measuring technique was developed which allows to measure three complanary lattice parameters of the unit cell (the amounts of two basic vectors and the inclosed angle between them) in one state of adjustment, in one measuring procedure, and at one position of the crystal. In the case of this extended BOND-method it is necessary to measure the angles between four different crystal positions (GROSSWIG et al. 1985), in which the Bragg angles are determined for two different asymmetric (or symmetric/asymmetric, respectively) atomic planes

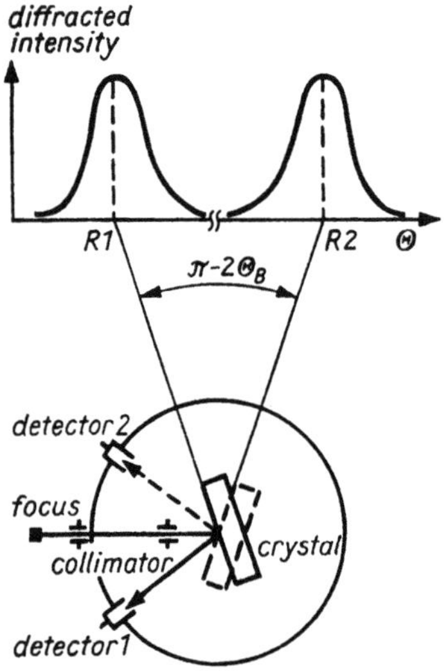

Fig. 1. Measuring principle of BOND-method

in two at time different reflection positions *hkl*. In the same procedure also the angle between the two used net planes can be measured. In the measuring procedure it is necessary to choose a geometry in which the zone axis of two used net planes is directed parallel to a suitable coordinate axis of the unit cell and this zone axis must be adjusted in a parallel direction to the goniometer axis.

The spatial deformation of unit cell has been determined by the BOND-method with the automatic measuring equipment JARD ("Jenaer Automatisches Roentgen-Diffraktometer", GROSSWIG et al. 1983). The attainable measuring accuracy of equipment in the case of optimal measuring conditions was found to be $|\delta a_3/a_3| = 2 \cdot 10^{-7}$ for the amount of basic vector perpendicular to the surface and $|\delta\gamma| = 2 \cdot 10^{-6}$ for the inclosed angle between two different reflection planes *hkl* (asymmetric/asymmetric or symmetric/asymmetric). A single measurement only provides information on a local region of sample because the incident X-ray beam had a horizontal extension of 0.28 mm and a vertical extension of 0.75 mm. The local measuring region follows from the extension of the plane of intersection between the X-ray beam and the surface of crystal. A scanning of the sample allows to get information on the distribution of lattice parameters along the surface of the crystal.

2. Specifics of measurements in LiF

The LiF crystals have been grown by the Bridgman-Stockbarger technique in vacuum with an as-grown diameter of 200 mm. On advanced growth conditions it is possible to get LiF samples with block desorientation angles of a few angular minutes or angular seconds only (DRESSLER, WEHRHAN). By preliminary examinations with the two-crystal diffractometry and topography the samples chosen from different places of the as-grown crystal are characterized regarding their block structure. The measuring surface was a cleavage face (001) (Fig. 2a). We used scanning steps of 0.5 mm between the different measuring points parallel to the face (001). The lattice parameters can be measured in a thin subsurface layer only (extinction depths for the used X-ray reflections: 4–5 µm, absorption depths: 30–40 µm). In this thin subsurface layer elastooptical measurements were also made in corresponding directions (Fig. 2a).

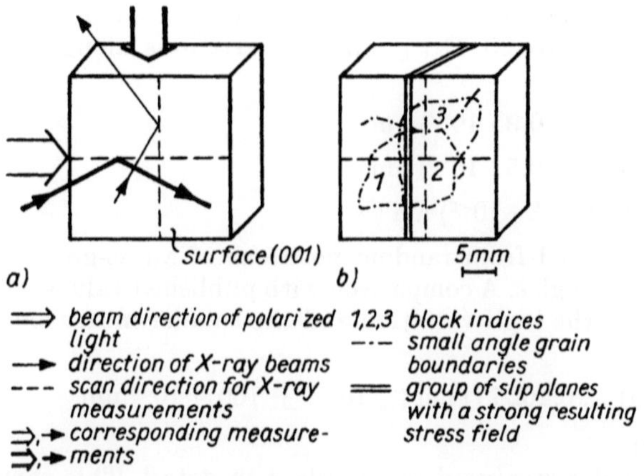

Fig. 2. Examined cleavage sample {100} of LiF (in a schematic manner). (a) measuring directions; (b) survey of block structure

Table 1
Parameters of $CuK_{\alpha 1}$ reflections at (001) surfaces of LiF crystals (with $\lambda(CuK_{\alpha 1}) = 0.154056\,2$ nm)

Type of reflection *hkl*	Bragg angle θ_B	length of horizontal measuring region	height of measuring region
004	49.94°	0.6 mm	0.75 mm
024	58.83°	0.9 mm	

Table 1 shows the parameters of used X-ray reflections. We already get the complete information on the variation of lattice parameters in the reflection plane (distortion of the unit cell in the section between reflection plane and the crystal) from the asymmetric reflections of type 024. The additional measurements of reflection 004 (symmetric) were important for the testing of measuring reliability (concerning the lattice parameter a_3).

3. Results of precise measurement of LiF lattice parameters

The above mentioned measuring accuracy of $|\delta a_3/a_3| = 2 \cdot 10^{-7}$ (or $|\delta a_3| = 0.12 \times 10^{-6}$ nm, respectively; measurement in Si, Grosswig et al. 1983) could not be attained for LiF because the Bragg angle considerably differs from 90° ($|\delta a_3| \approx \cot \theta_B$). A measuring accuracy like this has not been necessary due to a higher fluctuation of lattice parameters in LiF crystals. We got the following variations of lattice parameter a_3 (a_3 perpendicular to the surface of type (001)):

$|\delta a_3| = 0.5 \cdot 10^{-6}$ nm in a sequence of measurements at the same position, also in the cases of transition from the reflection 004 (symmetric) to 024 (asymmetric)

$|\delta a_3| = 1.3 \cdot 10^{-6}$ nm as middle variation in a single sample (average value for the three examined samples)

$|\delta a_3| = 10 \cdot 10^{-6}$ nm as maximal difference of average values from crystal to crystal.

The following average values of lattice constants of three single samples are corrected dates with reference to 20 °C (± middle variations along the surfaces of samples):

Sample 1: $a_3 = (0.4025758 \pm 0.9 \cdot 10^{-6})$ nm

Sample 2: $a_3 = (0.4025852 \pm 0.5 \cdot 10^{-6})$ nm

Sample 3: $a_3 = (0.4025862 \pm 3 \cdot 10^{-6})$ nm

The samples were separated from random positions of an as-grown crystal with small block desorientation angles. A comparison with published values is only possible in a rough manner due to the high variation and the smaller measuring accuracy of published values:

$$a = (0.40201 \dots 0.40277)\ \text{nm}\ (\pm 10 \dots \pm 100) \cdot 10^{-6}\ \text{nm}$$

(Landolt-Boernstein).

Table 2 shows the results measured at sample 3 in detail. This sample had the largest variation along the measuring surface in comparation to the other both samples. The upper index k of the parameters ${}^k a_i$ characterizes different blocks (survey of block structure in Figure 2b).

In Figure 3 the resulting distorsions of unit cell according to Table 2 are shown in a schematic manner. The values of shearing angles ${}^k\gamma_{ji}$ amount to a few angular seconds but the "inflexible" block desorientations amount to a considerably larger value of some 10 angular seconds (examined by two-crystal diffractometry).

Further on the measuring results correspond with the relation: a_1 & $a_2 > a_3$ for the amounts a_i of basic vectors (Tab. 2) along the whole examined region in both

Table 2

Variation of the amounts ${}^k a_i$ of basic vectors and shearing angles ${}^k\gamma_{ji}$ in the examined blocks k of LiF samples (survey of block structure in Figure 2b)

plane (y, z)	plane (x, z)
${}^2a_2 = 0.4025972$ nm	${}^2a_1 = 0.4026005$ nm
${}^2a_3 = 0.4025874$ nm	${}^2a_3 = 0.4025901$ nm
${}^1a_2 = 0.4025916$ nm	${}^3a_1 = 0.4026005$ nm
${}^3a_3 = 0.4025833$ nm	${}^3a_3 = 0.4025889$ nm
${}^2\gamma_{23} = 0$	${}^2\gamma_{13} = +2'' \pm 1''$
${}^1\gamma_{23} = 0$	${}^3\gamma_{13} = +1'' \pm 1''$
${}^2a_2 - {}^2a_3 \approx 1 \cdot 10^{-5}$	${}^2a_1 - {}^2a_3 \approx 1 \cdot 10^{-5}$
${}^1a_2 - {}^1a_3 \approx 1 \cdot 10^{-5}$	${}^3a_1 - {}^3a_3 \approx 1 \cdot 10^{-5}$
${}^2a_2 - {}^1a_2 \approx 5 \cdot 10^{-6}$	${}^2a_1 - {}^3a_1 \approx 0$
${}^2a_3 - {}^1a_3 \approx 5 \cdot 10^{-6}$	${}^2a_3 - {}^3a_3 \approx 0$

${}^k a_i$ amount of basic vector i in block k

${}^k\gamma_{ij}$ shearing angle between the basic vectors a_i and a_j in block k

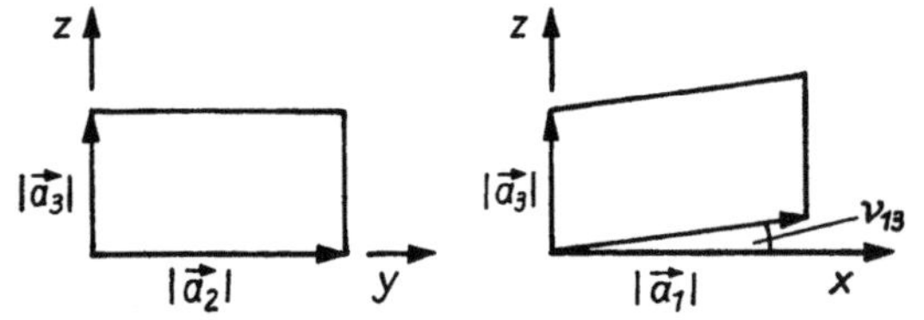

Fig. 3. Distortion of the unit cell (in a schematic manner)

measuring planes. In the plane (y, z) any shearing stresses could not be proved but all blocks show shearing stresses in the plane (x, z).

4. Comparison with elastooptic measurements

In the above-mentioned elastooptic observation directions (Fig. 2a) we can determine the same components of deformation corresponding to the X-ray results ($\varepsilon_{yy} - \varepsilon_{zz}$, or $\varepsilon_{xx} - \varepsilon_{zz}$, respectively). The measuring regions in the depths of crystal especially are not identical, however, but we tried to get information on the optical phase difference also near the surface (001). We got the following additional results:

the difference of lattice parameters between block 1 and 2 (Tab. 2) is likely caused by strong slip planes near the small angle grain boundary (Fig. 2b). The difference of lattice parameter in slip planes may be explained on the basis of elastomechanic calculations. In the same way the modulation of lattice parameters is probably caused in the interior of blocks in which we observed weak slip planes in the elastooptic images (an orthoscopic light path with crossed polarizers was used). The signs of elastooptic results (compression or dilatation, respectively) on both sides of the strong slip plane also correspond with the signs of the differences of lattice parameters.

The elastooptical determined deformations $\varepsilon_{yy} - \varepsilon_{zz}$ or $\varepsilon_{xx} - \varepsilon_{zz}$ respectively, near the surface (001) amount to values $\lessapprox 50\%$ in comparison to the X-ray results. As reason for this we suppose that the difference in the optical and X-ray values results from the different lateral extension of measuring regions (0.75 mm in the X-ray case and the whole length of sample in the optical case). Special information of the elastooptical measurements are given by GRIEBNER.

The investigations were supported by VEB Carl Zeiss JENA. The authors wish to thank Dipl.-Krist. R. Gütt (Forschungszentrum des Kombinates VEB Carl Zeiss JENA) for providing and preparing the crystals.

References

BOND, W. L.: Acta Cryst. **13**, 815 (1960)
DRESSLER, L., WEHRHAN, O.: Cryst. Res. Technol. 18, 1595 (1983)
GRIEBNER, U.: Thesis of diploma, Jena 1986
GROSSWIG, S., HÄRTWIG, J., ALTER, U., CRISTOPH, A.: Cryst. Res. Technol. 18, 501 (1983)
GROSSWIG, S., JÄCKEL, K.-H., KITTNER, R., DIETRICH, B., SCHELLENBERGER, U.: Cryst. Res. Technol. **20**, 1093 (1985)
LANDOLT/BÖRNSTEIN: Neue Serie III/7a, Berlin-Heidelberg-New York 1973

(Received April 21, 1987)

Authors' addresses:

Dr. sc. L. DRESSLER, Dr. R. KITTNER
Friedrich-Schiller-Universität, Sektion Physik
Max-Wien-Platz 1
Jena
DDR-6900

Dipl.-Phys. U. GRIEBNER
Akademie der Wissenschaften der DDR, Institut für Mechanik
Uhlandstr. 4
Karl-Marx-Stadt
DDR-9061

Cryst. Res. Technol.	22	1987	11	1436

Book Review

P. VILLARS, L. D. CALVERT

Pearson's Handbook of Crystallographic Data for Intermetallic Phases

American Society for Metals. Metals Park. Ohio. 1986. Vols. 1–3. 3258 pp, US $ 495.00
ISBN 0-87170-217-7

Das populäre Nachschlagewerk für Gitterkonstanten von WILLIAM B. PEARSON aus den Jahren 1958 und 1967, das auf metallische Elemente, Mischkristalle und Verbindungen spezialisiert ist, hat mit dem vorliegenden stark erweiterten Handbuch die langersehnte Fortsetzung bzw. Überarbeitung erfahren. Dabei ist nicht nur die neuere Literatur bis Mitte 1983 aufgearbeitet und in den Gesamtrahmen des Zeitraums 1913—1983 eingeordnet worden. Berücksichtigt wurden nun alle Phasen beliebiger Komponentenzahl mit Ausnahme von Halogeniden und ternären bzw. quaternären Oxiden. (Daß damit ausgerechnet die nach Redaktionsschluß interessant gewordenen Hochtemperatursupraleiter ausgeschlossen werden, mag auch die Autoren schmerzlich stimmen).

Ein besonderer Vorzug des PEARSONschen Werkes gegenüber anderen Gitterkonstantensammlungen ist die Ordnung nach Strukturtypen. Auch dieser wird in der vorliegenden Darstellung weiter ausgebaut (rund 2200 Strukturtypen finden Berücksichtigung). Als neue Qualität bietet die Sammlung von VILLARS und CALVERT zu jedem Strukturtyp die Koordinaten der besetzten Atomlagen (einschl. Besetzungsfaktor). Damit ist die Strukturinformation komplett. Natürlich können auch die Computerkunststücke der Autoren nichts an dem Umstand ändern, daß viele Lagekoordinaten der rund 22000 erfaßten Phasen bisher nicht oder nur ungenau bestimmt worden sind. Auch wenn es über den Strukturtypbegriff und die Nomenklatur zu dessen eindeutiger Kennzeichnung noch immer keine allgemeine Übereinkunft gibt, so erleichtert doch der eingeschlagene Weg des vorliegenden Werkes die praktische Arbeit ganz wesentlich. Daß die Strukturtypflut auch zu Redundanz führt, wird schon aus dem beigefügten Korrekturverzeichnis deutlich.

Der Leser kann dem Werk die Strukturdaten einer Verbindung entnehmen, deren Bruttoformel bekannt ist (Bände 2 und 3). Die streng alphabetische Ordnung nach der Formel ermöglicht auch dem Laien schnellen Zugriff. Weiterhin kann bei bekannter Bruttoformel des Strukturprototyps neben der Beschreibung des Strukturtyps die Zusammenstellung aller Phasen entnommen werden, die mit dem Strukturtyp kristallisieren. Die Ordnung der Strukturtypen erfolgt nach dem PEARSON-Symbol, das in einer gesonderten Liste abgelesen werden kann. Im Falle von Mischkristallen findet man Verweise auf die Literatur, meist mit Angaben über Untersuchungen zum Phasendiagramm gekoppelt.

Die an sich sehr begrüßenswerte Angabe der Lagekoordinaten ist allerdings in der vorliegenden Form noch mit Nachteilen behaftet. Abgesehen von der wohl unvermeidlichen hohen Druckfehlerquote an dieser Stelle ist die Darstellung durch Dezimalzahlen problematisch. Die freien Parameter sind in Band 1 an der Anzahl der Dezimalstellen herauszufinden, in Band 2 und 3 aber auch daran nicht mehr zu erkennen. Die Anzahl der unabhängigen freien Parameter ist weder aus Bd. 1 noch aus 2 und 3 zu gewinnen. Die Wahl des Koordinatenursprungs ist im übrigen nicht streng konform mit den Internationalen Tabellen (1983) erfolgt.

Die vorliegende Sammlung wertet rund 33000 Quellenangaben kritisch aus und verwendet für die Literaturzitate die ersten fünf Buchstaben des Autorennamens und eine eindeutig auflösbare Abkürzung der Publikation. Ohne elektronische Datenverarbeitung ist ein solches Vorhaben wohl nicht mehr zu realisieren. Aber trotz wachsender Verbreitung der Datenbankanschlüsse wird auf absehbare Zeit der Bedarf an gedruckten Datensammlungen bestehen bleiben.

Den Autoren ist der Dank des sehr breiten Nutzerkreises für die aufgewandte Mühe sicher.

P. PAUFLER

Cryst. Res. Technol.	22	1987	11	1437–1441

H. Neumann, E. Nowak, M. S. Omar

Sektion Physik and Sektion Chemie (WB Kristallographie),
Karl-Marx-Universität Leipzig, and
School of Physics, University of Bath, United Kingdom

High-temperature Heat Capacity, Thermodynamic Properties and Debye Temperature of $CuGe_2P_3$

The heat capacity at constant pressure of $CuGe_2P_3$ is measured in the temperature range from 180 to 550 K. Standard enthalpies and entropies relative to 298.15 K are calculated from the heat capacity data. The Debye temperature in the high-temperature limit is estimated to be about 710 K.

Die Wärmekapazität bei konstantem Druck wurde für $CuGe_2P_3$ im Temperaturbereich von 180 bis 550 K gemessen. Aus den gemessenen Wärmekapazitäten wurden Standard-enthalpien and Standardentropien relativ zu 298.15 K berechnet. Die Debyetemperatur für den Grenzfall hoher Temperaturen wurde zu etwa 710 K abgeschätzt.

1. Introduction

The ternary compound $CuGe_2P_3$ is a semiconductor crystallizing in the cubic sphalerite structure with a random distribution of the copper and germanium atoms on the cation sites (Folberth, Pfister; Goryunova et al.). The results of earlier investigations of $CuGe_2P_3$ regarding the phase diagram, microhardness, thermal conductivity and gap energy have been reviewed by Goryunova. Recently, there is a renewed interest in this compound and detailed studies of the elastic constants, thermal expansion behaviour, Grüneisen function and infrared lattice vibrations have been published in the literature (Hailing et al.; Neumann et al. 1987; Pamplin, Omar; Riede et al.). In the present work we report the heat capacity of $CuGe_2P_3$ at elevated temperatures. Standard enthalpies and entropies relative to 298.15 K and the high-temperature limit of the Debye temperature are determined from the measured heat capacity.

2. Experimental

The samples used in the experiments were prepared from $CuGe_2P_3$ single crystals grown by a modified Bridgman technique using a two-zone furnace (Hailing et al.). X-ray diffraction was used to confirm the structure and the lattice parameters of the compound and to ensure that the material is single phase. The heat capacity at constant pressure was measured in the temperature range from 180 to 550 K using a Setaram heat flow calorimeter model DSC 111. Details of the sample preparation, the measurement procedure and the calibration of the calorimeter have been reported elsewhere (Neumann et al. 1985b) and will not be repeated here. The estimated error in the measured heat capacities is about 1%.

3. Results and discussion

The measured temperature dependence of the molar heat capacity at constant pressure C_p of $CuGe_2P_3$ is shown in Figure 1. In complete analogy to the results reported for GaP (STRELCHENKO, LEBEDEV) which is the direct binary analog of $CuGe_2P_3$ we found

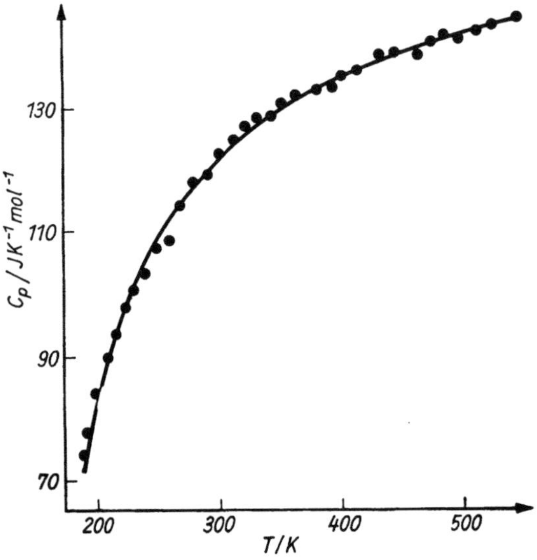

Fig. 1. Measured (points) and calculated with relation (1) (full curve) molar heat capacity at constant pressure of $CuGe_2P_3$

that within the error limits of the heat capacity determination and within the temperature range considered the experimental $C_p(T)$ values can be well described by the simple relation

$$C_p(T) = a - b/T^2 + cT\,. \tag{1}$$

A least-squares fit of relation (1) to the data points for $C_p(T)$ of Figure 1 yields $a = 149.7$ J/Kmol, $b = 2.644 \cdot 10^6$ J K/mol and $c = 5.6200 \cdot 10^{-3}$ J/K^2 mol. The resulting smoothed heat capacity values calculated with (1) are given in Table 1 for several fixed temperatures between 200 and 550 K. With relation (1) for $C_p(T)$ we

Table 1
Standard thermodynamic quantities of $CuGe_2P_3$ at $p_0 = 101\,325$ Pa

T (K)	C_p (J/K mol)	$H_m(T) - H_m(298.15)$ (kJ/mol)	$S_m(T) - S_m(298.15)$ (J/K mol)
200	84.7	−10.48	−42.14
250	108.8	−5.573	−20.35
300	122.0	0.225	0.753
350	130.1	6.541	20.21
400	135.4	13.19	37.95
450	139.1	20.05	54.12
500	141.9	27.08	68.93
550	144.0	34.24	82.56

have also simple relationships for the molar standard enthalpies and entropies relative to $T_0 = 298.15$ K which are defined as (LEVICH)

$$H_m(T) - H_m(T_0) = \int_{T_0}^{T} C_p(T)\,dT\,, \qquad S_m(T) - S_m(T_0) = \int_{T_0}^{T} \frac{C_p(T)}{T}\,dT\,. \tag{2}$$

The enthalpy and entropy increments calculated with relations (2) with (1) are compiled in Table 1, too.

With regard to the characteristic Debye temperature of $CuGe_2P_3$ the only experimental information available until now stems from the elastic constant measurements of HAILING et al. These authors calculated an elastic Debye temperature of $\theta^{el} = 429$ K which is expected to be very close to the low-temperature limit θ_0 of the thermal Debye temperature, similarly as in the case of the binary analog GaP where the corresponding Debye temperatures are equal to $\theta^{el} = 444$ K (HAILING et al.) and $\theta_0 = 436$ K (ABRAHAMS, HSU). In order to determine the Debye temperature θ_∞ for the high-temperature limit $T \to \infty$ one has to calculate and to analyse the apparent Debye temperature $\theta_D(T)$ defined as (BARRON et al.)

$$C_p(T) = 3NRF(\theta_D/T)\,, \tag{3}$$

where N is the number of atoms per molecular unit and $F(\theta_D/T)$ the Debye function. If lattice anharmonicity effects can be neglected the apparent Debye temperature is equal to the true Debye temperature of the compound. Now, using relation (3) and the heat capacity data from Figure 1 we find the temperature dependence of θ_D shown in Figure 2. We see that θ_D goes through a maximum at a temperature of

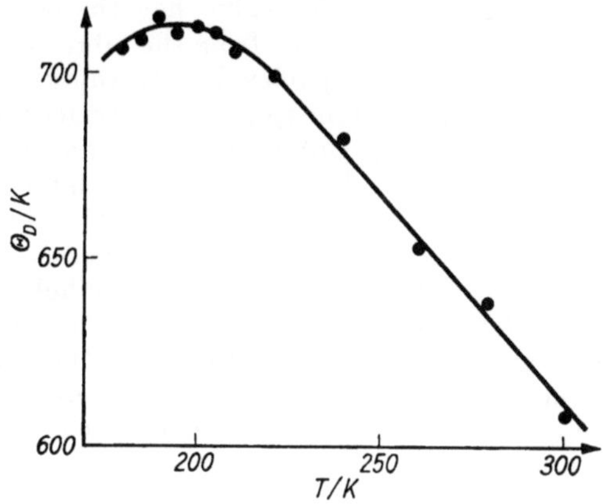

Fig. 2. Temperature dependence of the apparent Debye temperature of $CuGe_2P_3$

about 200 K and then decreases with increasing temperature. Since the decrease in θ_D can be ascribed to lattice anharmonicity effects (BARRON et al.) it is obvious that over the whole temperature range considered in the present study the contribution to C_p due to lattice anharmonicity cannot be neglected and, therefore, we are unable to deduce an accurate value for θ_∞ from the $\theta_D(T)$ curve of Figure 2. However, it can be expected that the maximum value of θ_D is approximately equal to or slightly lower than the high-temperature limit of the Debye temperature. Thus, it follows from Figure 2 that $\theta_\infty \gtrsim 710$ K in $CuGe_2P_3$. In order to get some evidence in favour of this supposition we can estimate the high-temperature limit of the Debye temperature from the vibrational characteristics of the compound. According to DOMB and

SALTER the high-temperature limit of the Debye temperature is given by the relation

$$\theta_\infty = \frac{h}{k}\left(\frac{5}{3}\mu_2\right)^{1/2} \tag{4}$$

where h is the Planck constant, k the Boltzmann constant and μ_2 the second moment of the phonon frequency spectrum of the compound. In principle, to evaluate μ_2 the knowledge of the phonon density of states is required which has not been measured for $CuGe_2P_3$ until now. However, if the general features of the phonon frequency spectrum and the interatomic force constants determining the vibrational characteristics are known a rough estimate of μ_2 can be made.

The interatomic force constants of $CuGe_2P_3$ have been determined from an analysis of the elastic constants (HAILING et al.) and the infrared lattice vibrational modes (RIEDE et al.) in good agreement with each other. Furthermore, in evaluating μ_2 we have to take into account that in complete analogy to the situation in the isostructural compound $CuSi_2P_3$ (NEUMANN et al. 1985a) there are at least nine optical modes in $CuGe_2P_3$ having approximately the same frequency and, therefore, giving rise to only one vibrational band in the infrared spectra with its centre of gravity at 375 cm^{-1} as observed experimentally (RIEDE et al.). Indeed, using the interatomic force constants from RIEDE et al., the bond ionicities from NEUMANN and the mode frequency relations from NEUMANN et al. (1985a) we find that the nine transverse optical mode frequencies are in the range from 372 to 391 cm^{-1} with its centre of gravity at 378 cm^{-1} which is in good agreement with the experimental value of 375 cm^{-1}. Thus, it seems to be justified to use this model to estimate the second moment of the phonon frequency spectrum. We find a value of $\mu_2 = 1.2 \cdot 10^{26}\ s^{-2}$ which if inserted in relation (4) yields a Debye temperature of $\theta_\infty = 680$ K. Comparing with the result deduced from the heat capacity data (Fig. 2) we see that this θ_∞ value has the correct order of magnitude, but that the Debye temperature estimated from the vibrational spectrum is about 30 K below that obtained from the heat capacity. Partly, this difference in the θ_∞ may be due to the fact that we have neglected the splitting between the longitudinal and transverse optical modes in evaluating μ_2. Unfortunately, these splittings have not been determined in $CuGe_2P_3$ because the infrared spectra are largely influenced by free carrier effects (RIEDE et al.). However, if we assume that the mode splittings are the same as in $CuSi_2P_3$ (NEUMANN et al. 1985a) the Debye temperature of $CuGe_2P_3$ becomes equal to $\theta_\infty = 690$ K which is indeed closer to but nevertheless lower than the value of about 710 K determined from the heat capacity. It can be argued that the mode splittings in $CuGe_2P_3$ are larger than those in $CuSi_2P_3$. On the other hand, it cannot be excluded that the high-temperature limit of the Debye temperature estimated from lattice vibration data is in general lower than that determined from the heat capacity because there is a similar difference in the corresponding θ_∞ values for GaP, the binary analog of $CuGe_2P_3$. In analysing existing experimental heat capacity data for GaP the high-temperature limit of the Debye temperature has been found to be $\theta_\infty = 560$ K (KAGAYA, SOMA). Since the lattice vibrational properties of GaP have been studied extensively by various authors (PÖDÖR and references cited therein) we can evaluate the second moment μ_2 of the phonon density of states and, thus, also θ_∞ rather accurately. The result is $\theta_\infty = 525$ K which is again clearly below the value determined from the heat capacity. At present, no explanation for this effect can be given.

References

ABRAHAMS, S. C., HSU, F. S. L.: J. Chem. Phys. **63**, 1162 (1975)

BARRON, T. H. K., BERG, W. T., MORRISON, J. A.: Proc. Roy. Soc. (London) A **242**, 478 (1957)

Domb, C., Salter, L.: Phil. Mag. **43**, 1083 (1952)
Folberth, O. G., Pfister, H.: Acta Cryst. **14**, 325 (1961)
Goryunova, N. A.: Halbleiter mit diamantähnlicher Struktur, Leipzig 1971
Goryunova, N. A., Sokolova, V. I., Bin-Si, T.: Dokl. Akad. Nauk SSSR **152**, 363 (1963)
Hailing, T., Saunders, G. A., Omar, M. S., Pamplin, B. R.: J. Phys. Chem. Solids **45**, 163 (1984)
Kagaya, H.-M., Soma, T.: phys. stat. sol. (b) **134**, K101 (1986)
Levich, V. G.: Introduction to Statistical Physics, Moscow 1954
Neumann, H.: Cryst. Res. Technol. **20**, K43 (1985)
Neumann, H., Kissinger, W., Hasoon, F. S., Pamplin, B. R., Sobotta, H., Riede, V.: phys. stat. sol. (b) **127**, K9 (1985a)
Neumann, H., Kühn, G., Möller, W.: Cryst. Res. Technol. **20**, 1225 (1985b)
Neumann, H., Nowak, E., Omar, M. S., Pamplin, B. R.: J. Less-Common Metals **132**, L13 (1987)
Pamplin, B. R., Omar, M. S.: Progr. Crystal Growth Character., **10**, 183 (1985)
Pödör, B.: phys. stat. sol. (b) **120**, 207 (1983)
Riede, V., Neumann, H., Sobotta, H., Kissinger, W., Omar, M. S., Pamplin, B. R.: phys. stat. sol. (b) **123**, K119 (1984)
Strelchenko, S. S., Lebedev, V. V.: A^3B^5 Compounds, Moscow 1984

(Received April 22, 1987)

Authors' addresses:

Dr. H. Neumann
Sektion Physik, Karl-Marx-Universität
Linnéstraße 5
Leipzig
DDR-7010

Dr. E. Nowak
Sektion Chemie, WB Kristallographie
Karl-Marx-Universität
Scharnhorststraße 20
Leipzig
DDR-7030

Dr. M. S. Omar
School of Physics, University of Bath
Claverton Down
Bath BA2 7AY
United Kingdom

Book Review

KESHRA SANGWAL

Etching of Crystals — Theory, Experiment, and Application

Defects in Solids, vol. 15

North-Holland, Amsterdam, Oxford, New York, Tokyo 1987, 520 Seiten, zahlreiche Abbildungen und Tabellen, Sach- und Autorenregister, Preis Dfl. 275.00, ISBN 0-444-87018-0

Die Monografie gibt einen umfassenden Überblick zum chemischen Ätzen. Schwerpunkte sind der Auflösungsprozeß als negatives Kristallwachstum und die Bildung von Versetzungsätzgruben.

Das Buch ist in vier Teile (11 Kapitel) gegliedert. Im ersten Abschnitt werden relativ knapp Defekte in Kristallen und die entsprechenden Nachweismethoden behandelt. Den Schwerpunkt des Buches bilden die Teile 2 und 3. Im zweiten Abschnitt (Kap. 3 bis 6) sind Kristallwachstum und -auflösung (u. a. Keimbildung, Wachstumskinetik, Wachstumsform), die Theorie der Auflösung und Ätzgrubenbildung und chemische Aspekte des Auflösungsprozesses beschrieben. Der dritte Teil (Kap. 7 bis 9) enthält Ausführungen zu experimentellen Ergebnissen und deren Interpretation (z.B. Auflösungskinetik und -mechanismen, Morphologie von Ätzgruben). Auswahlkriterien für Ätzlösungen (Versetzungs- und Polierätzmittel) und Anwendungsaspekte u. a. bei der plastischen Deformation und in der Halbleiterindustrie (Strukturierung usw.) sind Gegenstand des 4. Abschnittes (Kap. 10 und 11). Im Anhang werden tabellarisch Versetzungs- und Polierätzer aufgeführt.

Mit dem Buch von SANGWAL liegt nunmehr eine in sich geschlossene Darstellung zum chemischen Ätzen vor. Theorie, Experiment und Anwendungsaspekte werden behandelt. Man findet stofflich gesehen eine Übersicht zum chemischen Ätzen der verschiedenen Kristallarten (Isolatoren, Halbleiter, Metalle), wobei selbst in dieser umfangreichen Arbeit z.B. bezüglich der Halbleiter nicht alle Aspekte ausführlich beschrieben werden können (u. a. nur eine Seite zum Lichtätzen), sondern auf Reviewartikel verwiesen werden muß. Die mehr als 900 Literaturzitate sowie zahlreiche Abbildungen und Tabellen unterstreichen den Überblickscharakter der Monografie.

Das Buch schließt eine Lücke in der Literatur zum Ätzen und ist allen zu empfehlen, die sich mit der Ätzthematik im engeren Sinne, aber darüber hinaus auch mit Problemen des Kristallwachstums und der Kristallbaufehler beschäftigen.

B. SCHUMANN

Cryst. Res. Technol.	22	1987	11	K173–K176

Short Notes

V. L. SEDOV

Physics Department, Moscow State University, Moscow, USSR

Some Physical Properties of Impurities in the Anderson Model[1])

This paper presents the basic physical properties of impurity atoms that follow from the general form of exact solutions of the nondegenerate Anderson model. These solutions are used to study the interaction of positrons with magnetic impurities via the field of virtual plasmons.

1. Basic properties

In the Anderson theory, which is based on the Hartree-Fock approximation, an impurity atom of a transition element in a nonmagnetic matrix can be in two different physical states; one magnetic and the other nonmagnetic. According to the theory, the spin magnetic moment of an impurity atom is given by the formula (ANDERSON)

$$\langle m_z \rangle = (n_+ - n_-)\, \mu_B \,, \tag{1}$$

where n_+ and n_- are the average occupation numbers of the localized d-level with spins the directions of which are specified by the + and − subscripts and μ_B is the Bohr magneton. Although $\langle m_z \rangle$ described by Eq. (1) is the average value of the projection of the atomic magnetic moment on the z-axis, in the Anderson theory it is regarded as a quantum number. Experiments aimed at investigating the internal magnetic fields H_{int} on magnetic impurity nuclei and the spin relaxation at low temperatures have not corroborated this conclusion (CAMPBELL et al.; LOEWENHAUPT, JUST). In the absence of an external magnetic field, the internal magnetic fields on the impurity nuclei are zero, although according to Eq. (1) H_{int} must have a non-zero value. Thus, the magnetic impurities are in a state in which (LOWENSTEIN, SEDOV)

$$\langle m_z \rangle = 0 \,, \qquad \langle \overline{m}^2 \rangle \neq 0 \,. \tag{2}$$

In the simplest case of non-degenerate model, the Anderson Hamiltonian has the form

$$H = \sum_{\bar{k},\sigma} \varepsilon_{\bar{k}} \overset{+}{a}_{\bar{k}\sigma} a_{\bar{k}\sigma} + \sum_{\sigma} \varepsilon^0_{\rm d} \overset{+}{a}_{{\rm d}\sigma} a_{{\rm d}\sigma} + v \overset{+}{a}_{{\rm d}\sigma} a_{{\rm d}\sigma} \overset{+}{a}_{{\rm d},-\sigma} a_{{\rm d},-\sigma} + \\ + v \sum_{\bar{k},\sigma} (\overset{+}{a}_{\bar{k},\sigma} a_{{\rm d}\sigma} + \overset{+}{a}_{{\rm d}\sigma} a_{\bar{k}\sigma}) \,. \tag{3}$$

The notation here is quite common. The symbols $\overset{+}{a}_{\bar{k}\sigma}$ and $a_{\bar{k}\sigma}$ stand for the operators representing the creation and annihilation of conduction electrons in a state with

[1]) Contributed paper presented during the European Meeting on Positron Studies of Defects, March 23–27, 1987, Wernigerode, G.D.R.

wave vector $\boldsymbol{k}$ and a spin the direction of which is specified by subscript σ. The operators $a_{d\sigma}^{+}$ and $a_{d\sigma}$ have a similar meaning for an electron on the localized d, -impurity level. Also, ε_d^0 is the energy of the localized d-electron in the field of an impurity ion. v is the coupling constant of the intra-atomic interaction, and v is the transition integral between the s-states and the d-state. In the absence of the s–d interaction, i.e., $v = 0$, the system under consideration consists of two non-interacting parts, namely, an isolated impurity atom and an degenerate electron gas. There are three different energy states of the impurity atom in such a system for the degenerate electron gas in the ground state. The state with the energy E_1^0 corresponds to the case where the impurity atom contains a single electron. The energy is

$$E_1^0 = E^0 + \varepsilon_d^0 , \tag{4}$$

where E^0 is the energy of the ground state of the degenerate electron gas consisting of N electrons, and ε_d^0 is the energy of the localized d-level. In the second state with an energy E_2^0, the impurity atom contains two electrons on the d-level, and

$$E_2^0 = E^0 - \varepsilon_F + 2\varepsilon_d^0 + v , \tag{5}$$

here ε_F is the Fermi energy, and v is the coupling constant of the intra-atomic interaction. The third state with an energy E_3^0 corresponds to the case where both d-levels are vacant

$$E_3^0 = E^0 + \varepsilon_F . \tag{6}$$

When the s–d interaction is switched on, these three states mix and electron-hole pairs are excited in the degenerate electron gas. Due to such processes, the wave function of the entire system can be written as

$$\Phi = A_1(a_{d\sigma}^{+}\Psi_1 + a_{d,-\sigma}^{+}\Psi_1') + A_2 a_{d\sigma}^{+} a_{d,-\sigma}^{+} + A_3\Psi_3 , \tag{7}$$

where conditions (2) have been allowed for. The functions Ψ_1, Ψ_1', Ψ_2, and Ψ_3 describe the states of the degenerate electron gas interacting with the impurity atom; A_1, A_2, and A_3 are constant coefficients. This wave function written in form (7) is a superposition of the impurity states corresponding to the occupation numbers of the localized d-level equal to 1, 2, and 0. Formula (7) implies that there are three solutions, Φ_1, Φ_2, and Φ_3 of the Schrödinger equation with the Anderson Hamiltonian that differ only in values of the coefficients A_1, A_2, and A_3 (SEDOV). Thus, there can be three different states of the impurity atom. The physical meaning of these solutions to the Schrödinger equation is determined by the three states of the impurity atom in the absence of interaction (Eqs (4)–(6)). In the Anderson model, which is based on the Hartree-Fock approximation, an impurity atom can have only two states. One is magnetic with $\langle m_z \rangle \neq 0$ and the other is non-magnetic with $\langle m_z \rangle = 0$. For exact solutions of the form (7), one can also speak of different magnetic states of the impurity atom. However, in the non-magnetic state the coefficient $A_{1\,\text{non-mag.}}$ in (7), which determines $\langle \overline{m}^2 \rangle$ is not zero. One can only assert that

$$|A_{1\,\text{non-mag.}}|^2 < |A_{1\,\text{mag.}}|^2 . \tag{8}$$

2. The impurity-positron interaction

The density of the spatial positron distribution $\varrho(\bar{x})$ is determined by the positron Green function $G_p(x, x; t - t')$:

$$\varrho(\bar{x}) = \lim_{t' \to t+0} G_p(\bar{x}, \bar{x}; t - t') . \tag{9}$$

The sought expression for G_p that allows for the interaction of a positron with the impurity atom can be found by employing the Dyson equation

$$G_p(\bar{x}, \bar{x}'; t - t') = G_p^0(\bar{x} - \bar{x}'; t - t') + \\ + \int G_p^0(\bar{x} - \bar{\xi}; t - t_\xi)\, M(\bar{\xi}, \bar{\eta}; t_\xi - t_\eta)\, G_p(\bar{\eta}, \bar{x}'; t_\eta - t')\, d^4\xi\, d^4\eta\,, \qquad (10)$$

where G_p^0 is the Green function of a free positron. The simplest expression for the mass operator M that allows for long-range positron-electron interaction is

$$M(\bar{\xi}, \bar{\eta}; t_\xi - t_\eta) = \int v_{ep}(\xi - \alpha)\, G_e(\bar{\alpha}, \bar{\beta}; t_\alpha - t_\beta) \times \\ \times G_e(\bar{\beta}, \bar{\alpha}; t_\beta - t_\alpha)\, v_{ep}(\beta - \eta)\, G_p^0(\xi - \eta)\, d^4\alpha\, d^4\beta\,, \qquad (11)$$

here v_{ep} — effective potential positron-electron interaction. If the impurity atom interacts weakly with the electron gas, i.e., the d-level width 2Γ is much smaller than v and ε_F only the diagonal part G_e^{dd} corresponding to the localized d-level can be allowed for in the expression for the electron Green function G_e in (11). In this approximation, the wave functions (7), which describe the states of an impurity in a degenerate electron gas, have the form

$$\Phi_1 \approx \frac{1}{\sqrt{\alpha}}\,(a_{d\sigma}^+ \Psi_1 + a_{d,-\sigma}^+ \Psi_1'); \qquad \Phi_2 \approx a_{d\sigma}^+ a_{d,-\sigma}^+ \Psi_2; \qquad \Phi_3 \approx \Psi_3\,. \qquad (12)$$

These wave functions correspond to energy values defined via (4)—(6).

Let us consider the case where the ground state is the magnetic state of the impurity, i.e., a state with the wave function Φ_1. In the approximation we are using here, the delay and forestall parts of the Green function, $G_{e+}^{dd,\sigma}$ and $G_{e-}^{dd,\sigma}$, have the following simple form:

$$G_{e\pm}^{dd,\sigma}(\bar{x}_1, \bar{x}_2; \omega) = \frac{1}{4\pi}\,\frac{\varphi_{d\sigma}(\bar{x}_1)\, \varphi_{d\sigma}^*(\bar{x}_2)}{\omega - \varepsilon_\pm \pm i\delta}\,, \qquad (13)$$

where $\varphi_{d\sigma}$ is the atomic wave function of the d, σ-level, and

$$\varepsilon_+ = E_2^0(N + 1) - E_1^0(N) = \varepsilon_d^0 + v\,, \\ \varepsilon_- = E_1^0(N) - E_3^0(N - 1) = \varepsilon_d^0\,. \qquad (14)$$

Only the part corresponding to long-range interaction will be taken into account in the expression for the effective potential v_{ep}. The Fourier transform of this quantity is

$$v_{ep}(\bar{q}, \omega) = \frac{4\pi\, e^2}{q^2}\,\frac{\omega^2}{\omega^2 - \omega_0^2}\,, \qquad (15)$$

where ω_0 is the frequency of plasma oscillations in the electron gas. In view of the above remarks, the expression for $\varrho(\bar{x})$ can be written in the form

$$\varrho(\bar{x}) = \varrho^0 + \varrho'(\bar{x}) \qquad (16)$$

here $\varrho^0 = V^{-1}$ (V is the sample volume) and $\varrho'(\bar{x})$ is the positron distribution density near the impurity given by the second term on the right-hand side of (10) in which G_p is replaced with G_p^0. In explicit form,

$$\varrho'(\bar{x}) = \frac{e^4 v}{(2\pi)^4\, V\omega_0[\omega_0 + v]^2} \Bigg\{ \int\limits_{|\bar{q}| \le p_c;\ |\bar{p} - \bar{q}| \le p_c} [Q^*(\bar{q})\, Q^*(\bar{p} - \bar{q})\, e^{-i\bar{p}\bar{x}} + \\ + Q(\bar{q})\, Q(\bar{p} - \bar{q})\, e^{i\bar{p}\bar{x}}]\, p^{-2}(\bar{p} - \bar{q})^{-2}\, d^3p\, d^3q - \\ - \int\limits_{|\bar{p}_1| \le p_c;\ |\bar{p}_2| \le p_c} Q(\bar{p}_1)\, Q^*(\bar{p}_2)\, e^{i(\bar{p}_1 - \bar{p}_2)\bar{x}}\, p_1^{-2} p_2^{-2}\, d^3p_1\, d^2p_2 \Bigg\}\,, \qquad (17)$$

where

$$Q_d(\bar{p}) = \int e^{-i\bar{p}\bar{\xi}} \, |\varphi_d(\bar{\xi})|^2 \, d^3\xi \tag{18}$$

and we have omitted terms that contain a factor of the order of ε_c/ω_0 with ε_c the kinetic energy of a positron the momentum of which is equal to the critical plasmon momentum p_c. At values of x for which $p_c x < 1$ (x is the distance from the center of the impurity) $\varrho'(\bar{x})$ is a constant, whereas for distance that satisfy the condition $p_c x \gg 1$ this function is proportional to x^{-1}. At $p_c x \sim 1$, and $v \sim \omega_0$ the ratio ϱ'/ϱ^0 is of the order 0.1. Thus, a certain increase of positron density occurs in the vicinity of the impurity, at distances exceeding the characteristic action radius of the screening Coulomb potential.

References

Anderson, P. W.: Phys. Rev. **124**, 41 (1961)
Campbell, S. J., Clark, P. E., Hicks, T. J.: Solid State Commun. **24**, 791 (1977)
Loewenhaupt, M., Just, W.: Phys. Lett. **53A**, 305 (1975)
Lowenstein, J. H.: Phys. Rev. **B29**, 4120 (1984)
Sedov, V. L.: Pis'ma v JETF **36**, 128 (1982)

(Received April 23, 1987)

Authors' address

V. L. Sedov
Physics-Department, Moscow University
USSR-119899 Moscow

Cryst. Res. Technol.	22	1987	11	K177—K179

M. Ederhof, W. Lühr-Tanck, A. Sager, Th. Kurschat, Th. Hehenkamp

Institut für Metallphysik, Universität Göttingen and SFB 126, Göttingen/Clausthal

New High Precision Lifetime Measurements in Pure Copper and a 0.3 at.% AgSb Alloy with a BaF_2 Lifetime Spectrometer[1])

The temperature dependence of positron annihilation in copper and a 0.3 at.% AgSb alloy has been investigated by measuring the positron lifetime. In both cases the vacancy lifetime is nearly independent on temperature while the bulk lifetime shows a rapid increase after the onset of detectable vacancies. This increase can be explained by trapping during thermalization with an enhanced trapping rate.

1. Introduction

It is well established that positron annihilation studies can give information about defects in metals. In previous papers we determined the vacancy formation energy in copper and silver by using the lifetime technique (Lühr-Tanck et al.; Hehenkamp et al.). Recently Schäfer et al. presented lifetime measurements on copper showing a vacancy lifetime of about 180 ps for temperatures of about 600 K which decreases with increasing temperature, reaching at least the "normal" vacancy lifetime in copper of about 160 ps. Our previous lifetime measurements on copper have not shown such a behaviour. So we carried out new measurements with an increased accuracy using a new BaF_2 spectrometer to clarify this point.

2. Experimental

The specimen preparation is described in earlier papers (Lühr-Tanck et al.; Hehenkamp et al.). The equipment for the life-time-spectrometer is nearly the same as used earlier (Lühr-Tanck et al.), only the scintillators are now made of BaF_2 (35 mm diameter, 15 mm thickness) which are coupled to XP 2020 Q photomultipliers. The BaF_2-crystals are polished and covered by teflon tape. In this way we could observe a real time resolution (^{60}Co, Na-wind.) of 240 ps together with count rates of about 70 cps. A typical lifetime spectrum contained 6×10^6 counts.

3. Results and discussion

For the analysis of the lifetime spectra we employed the fit-program POSFIT (Kierkegaard, Eldrup). A source of about 1% intensity and a lifetime of 450 ps was sub-

[1]) Contributed paper presented during the European Meeting on Positron Studies of Defects, March 23—27, 1987, Wernigerode, G.D.R.

tracted from the spectra. Using the simple two-state trapping model two component fits were carried out if possible. The mean lifetime as well as the bulk lifetime are calculated by using the equations

$$\bar{\tau} = I_1\tau_1 + I_2\tau_2 \tag{1}$$

and

$$\lambda_f' = I_1\lambda_1 + I_2\lambda_2 \,. \tag{2}$$

Here I_1 and I_2 are the intensities of the two lifetimes τ_1 and τ_2, λ_1 and λ_2 are the corresponding decay constants. Figures 1 and 2 show the results for the lifetimes τ_1 and τ_2 in copper and a 0.3 at.% AgSb alloy

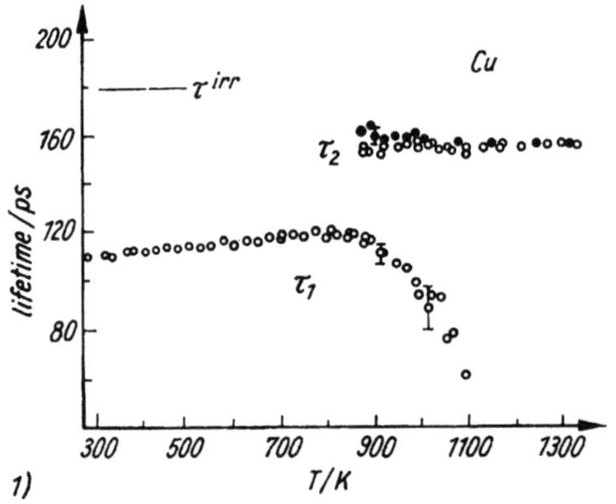

1)

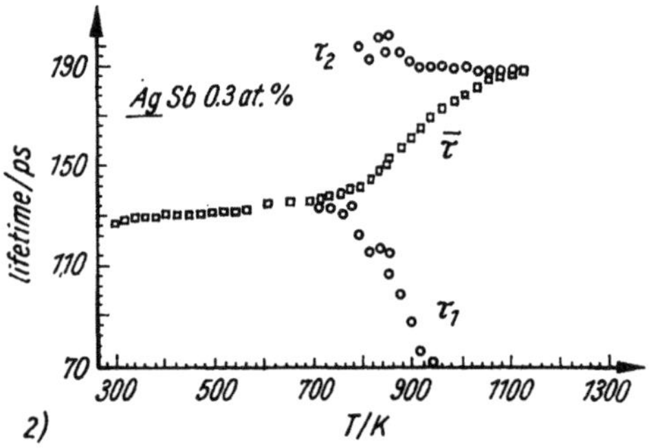

2)

Fig. 1. Lifetimes τ_1 and τ_2 for pure copper. The source term is subtracted. ○ old measurements (Hehenkamp et al.); ● new measurements (this work)
Fig. 2. Lifetimes τ_1 and τ_2 for a 0.3 at.% AgSb alloy. The source term is substracted

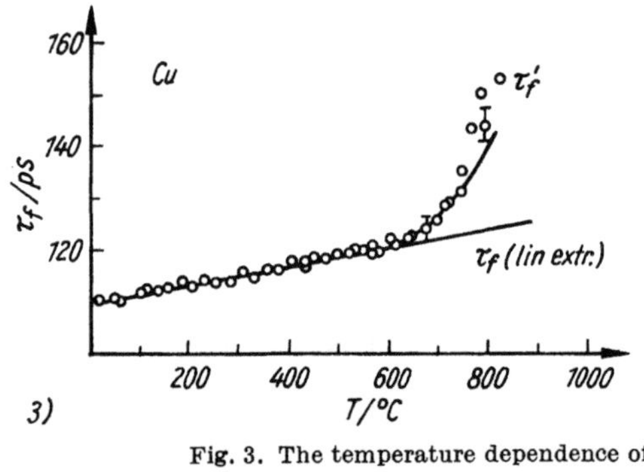

3)

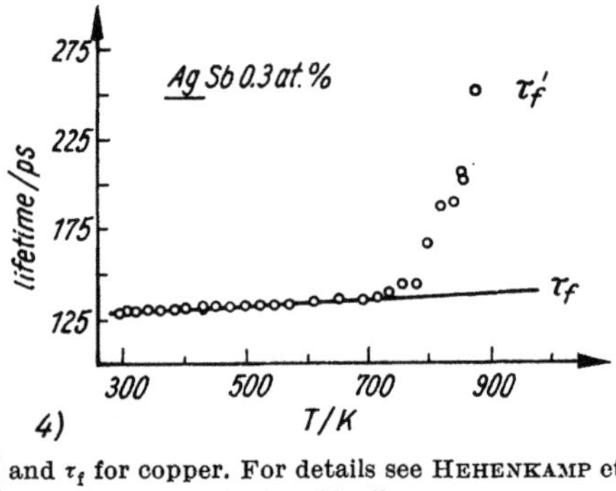

4)

Fig. 3. The temperature dependence of τ_f' and τ_f for copper. For details see Hehenkamp et al.
Fig. 4. The temperature dependence of τ_f' and τ_f for a 0.3 at.% AgSb alloy

It is obvious that our new measurements show nearly the same temperature-independent vacancy lifetime for copper as well as for the dilute silver alloy. So it seems very unlikely that the "real" vacancy lifetime especially for copper is in the order of 180 ps as proposed by Schäfer et al. We cannot find any effect of thermal detrapping with increasing temperature. We think that it must be an open question if the 180 ps observed after electron-irradiation reflects monovacancies or other more complex defects.

On the other hand the bulk lifetime τ_f' shows a strong temperature dependence as shown in Figures 3 and 4. Recently some studies about this point have been given by Nielsen et al. and McMullen and Stott. These authors conclude that a resonance trapping of non-thermal positrons is possible with the consequence of an enhanced trapping rate, more than an order of magnitude greater than for thermalized positrons.

In this way the dramatic increase of τ_f' can be understood, because these values are calculated by using the simple two-state trapping model.

As discussed in earlier papers (LÜHR-TANCK et al.; HEHENKAMP et al.), the determination of vacancy formation energies is not influenced by this resonance trapping if one uses the "$I_2(\lambda_1 - \lambda_2)$" method. For copper we find a value of $E_v^f = 1.28 \pm 0.04$ eV and for the 0.3 at.% AgSb alloy of $E_v^f = 1.10 \pm 0.04$ eV.

The authors would like to thank Dr. G. Kögel and Dr. Sperr for useful discussions for designing the BaF_2 lifetime spectrometer. We thank Deutsche Forschungsgemeinschaft and SFB 126 Göttingen for financial support.

References

HEHENKAMP, Th., KUSRCHAT, Th., LÜHR-TANCK, W.: J. Phys. **F 16**, 981 (1986)
KIERKEGAARD, P., ELDRUP, M.: Comput. Phys. Com. **3**, 240 (1972)
LÜHR-TANCK, W., KUSRCHAT, Th., HEHENKAMP, Th.: Phys. Rev. **B 31**, 2994 (1985)
McMULLEN, T., STOTT, M. J.: Phys. Rev. **B 34**, 8985 (1986)
NIELSEN, B., LYNN, K. G., CHEN, Y. C.: Phys. Rev. Let. **57**, 1789 (1986)
SCHÄFER, H. E., STUCK, W., BAUHART, F., BAUER, W., GUGELMEIER, R., SEEGER, A.: Proc. of the Conf. Vacancies and Interstitials in Metals and Alloys, Berlin 1986, in press

(Received April 23, 1987)

Authors' address:

M. EDERHOF, W. LÜHR-TANCK, A. SAGER, Th. KURSCHAT, Th. HEHENKAMP
Institut für Metallphysik
Hospitalstr. 3–5
D-3400 Göttingen
FR Germany

Cryst. Res. Technol.	22	1987	11	K180—K184

W. RUDZINSKA, CZ. SZYMANSKI, S. CHABIK, B. ROZENFELD

Institute of Physics, Pedagogical University, Opole and
Institute of Experimental Physics, University Wrocław, Wrocław

Investigation of the Effect of Temperature on the Positron Annihilation in Solid and Liquid Tin[1])

The temperature dependence of the counting rate, $F(T)$, has been measured in tin. A sudden increase in the counting rate, connected with the creation of vast defects acting as positron traps, has been observed on passing through the melting point. Both the $F(T)$ and angular correlation curves indicate that the defected areas decay with increasing temperature. The kinetics of this effect has been discussed and the vacancy formation energy, being equal to $E_V = 0.50 \pm 0.02$ eV, has been determined.

1. Introduction

The positron annihilation method (PAM) is widely used for determining the parameters of the crystal lattice defects. The equilibrium measurements are commonly used for determining the vacancy formation enthalpy and the non-equilibrium measurements for studying the kinetics of changes in the defect structure resulting from plastic deformation, irradiation etc. (WEST). However papers dealing with the PAM investigations of both the solid and liquid phase are scane. This is because of the experimental difficulties encountered when applying the PAM at high temperatures and especially in the existence region of the liquid phase of metals. This is the reason for which until now the results of PAM investigations of the liquid phase are available only for low-melting metals: Ga (BRANDT, WAUNG), Cd (SZYMAŃSKI et al.), Ga and Bi (SZYMAŃSKI et al. 1980), In and Ga (KISHIMOTO, TANIGAWA), Sn (MASCHER, PUFF). The interpretation of the results of measurements on these metals is also difficult because of complicated effects occurring in the prevacancy region (HERLACH et al.; SHAH, CATZ).

From the measurements performed on solid Sn it follows that: i) the changes in positron annihilation parameters in the prevacancy region differ from those observed for other low melting metals, ii) the changes of the positron annihilation parameters above the threshold temperature are relatively weak, iii) there is no saturation effect observed for other metals when approaching to the melting point.

In the present study the PAM measurements were performed for both the solid and liquid phases of tin with the aim to obtain a more complete information on the mechanism of interaction of positrons with crystal lattice defects in Sn.

2. Experiment

The temperature dependence of the counting rate, $F(T)$, has been measured for the cast (a) and formed in the measuring vessel (b) Sn samples in the temperature interval 20 to

[1]) Contributed paper presented during the European Meeting on Positron Studies of Defects, March 23—27, 1987, Wernigerode, G.D.R.

400 °C. The sample (a) was a strip cast from the same material. The samples (b) were investigated in the heating run in the temperature interval RT — 400 °C and in the cooling run in the temperature interval 400—200 °C. Next, after long-lasting soaking of the peak counting rate has been measured during continuous cooling to the room temperature. The measurements on the cast sample (a) where performed in both the heating and cooling runs in the interval RT — 190 °C. Additionally the angular correlation curves were mesured at 225 °C (solid state) and 235, 240, 250, and 370 °C (liquid state). The measurements were performed in the atmosphere oxygen-free dry argon, under normal pressure. During the measurement at each temperature, lasting 4 hours, 2.5 × 10^5 counts were accumulated. The measurements were performed by means of a two-channel annihilation spectrometer with long slits (Rozenfeld et al.).

3. Results and discussion

The temperature dependence of the counting rate, $F(T)$, measured in the heating run, shows a sudden increase at the melting temperature (T_M) and then, at 400 °C, a decrease to the room temperature value. During the cooling run the step-like change in the counting rate has been observed only on passing through the freezing point.

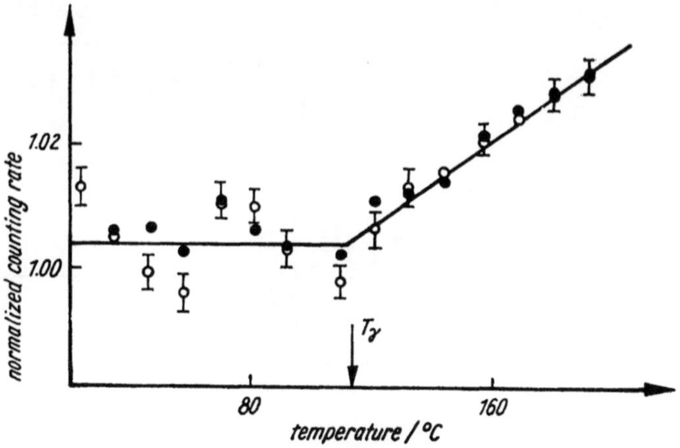

Fig. 1. Temperature dependence of the peak counting rate for the cast sample, ○ heating run, ● cooling run

The value of the coincidence counting rate at 400 °C, being equal to that measured at RT indicates, that the defects created during the melting process (probably the vacancy clusters or the ordered clusters of atoms) decompose due to the heating, and the true liquid phase structure is formed only after achieving the temperature of about 400 °C. The coincidence counting rate measured during long lasting (5 days) tempering of the sample at 250 °C (also just above T_M) was constant (Fig. 3), thus suggesting that at this temperature the system liquid metal-defect clusters is in thermodynamic equilibrium. During the cooling of the sample soaked at 400 °C the defect clusters acting as positron trapping centres are not formed anew, and therefore the coincidence counting rate in the whole liquid range remains to be constant (Fig. 2a). Thus, one concludes that in the temperature interval just above the melting point two different thermodynamic states can occur depending on the thermal history of the sample. An additional set of measurements in which the sample tempered at 400 °C was slowly cooled to the room temperature (Fig. 2b) gives the result similar to that shown in Figure 2a. This corroborates the conclusion drawn from the result

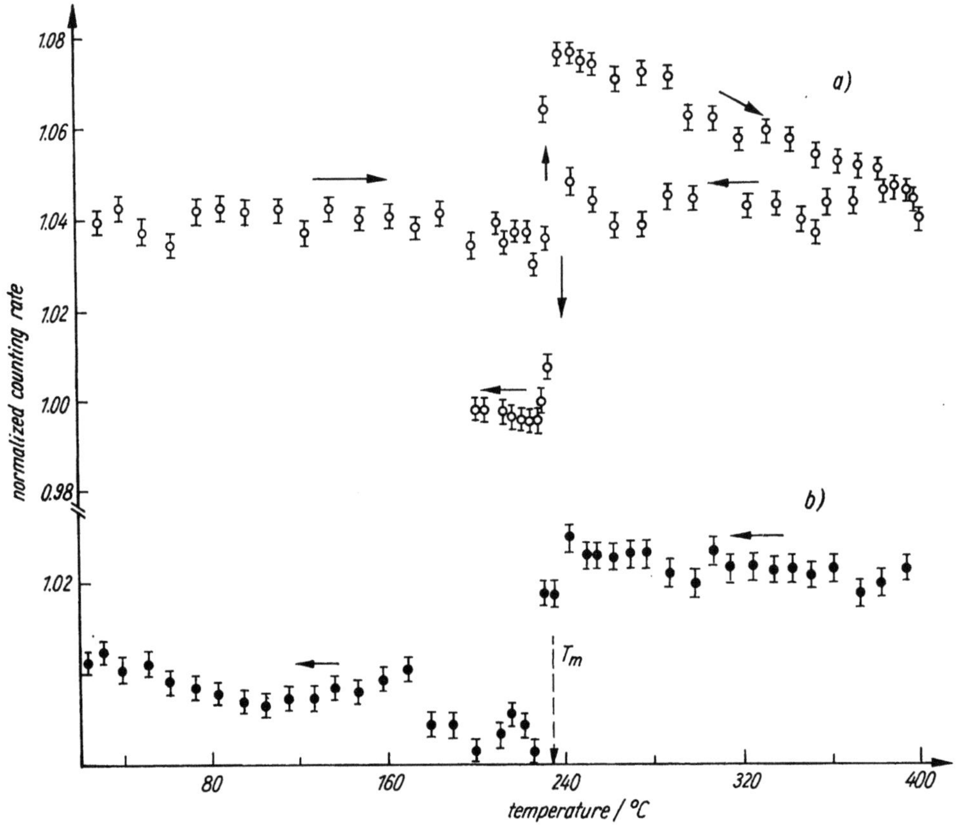

Fig. 2. Temperature dependence of the peak counting rate
a) heating and cooling run for the sample in measuring vessel; b) cooling run taken after reheating the sample at 400 °C

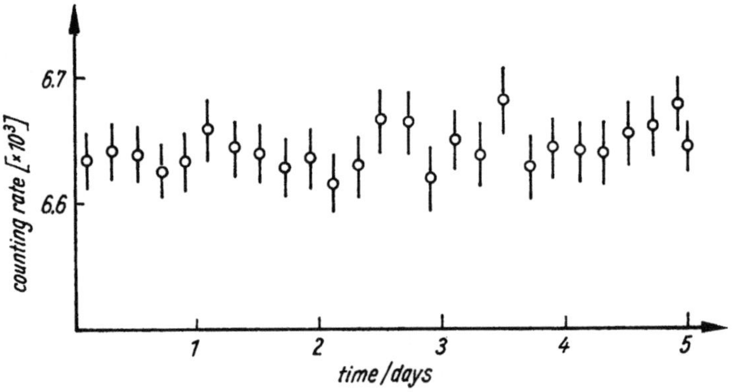

Fig. 3. Time dependence of the peak counting rate in 250 °C for the sample in measuring vessel

represented in Figure 2a according to which in the sample with the "true" liquid structure the defect clusters are not formed anew during the process of slow cooling (10^{-3} K · min^{-1}).

The most surprising feature of the result represented in Figure 2b is that the changes in the counting rate are opposite with respect to the S-shaped curve, commonly observed in the subsolidus region. One of the possible explanation of this is that in the slow cooling process of Sn a "supercooling" of the "true" liquid structure occurs and the crystal structure is reconstructed at temperatures of about 100 °C, at which a small increase in the peak counting rate is observed (Fig. 2b). The results of the measurements on the cast sample, represented in Figure 1, corroborate this supposition. As it can be seen from this figure, the course of the $F(T)$ curves measured in the heating and in the cooling run is nearly the same, and a characteristic increase in the counting rate occurs on both the curves above 120 °C. The threshold temperature $T_T = 114 \pm 5$ °C and the value of the vacancy formation energy, $E_v = 0.50 \pm 0.02$ eV, determined from these measurements agree with the results of other PAM measurements on solid Sn (Dedoussis et al.; McKenzie, Lichtenberger).

The results of the measurements of angular correlation curves agree well with the results discussed above. The values of the R parameter, characterizing the dimensions of the defects acting as positron traps, determined from angular correlation curves measured at different temperatures are collected in Table 1. As it can be seen from

Table 1
The values of the R parameter at different temperatures

phase	temperature, °C	R
solid liquid	225	0.26 ± 0.05
	235	1.86 ± 0.05
	240	2.35 ± 0.05
	250	2.41 ± 0.05
	330	0.92 ± 0.05

this table, the value of the R parameter suddenly increases on passing through the melting point, and then decreases above 330 °C. If the value of $R = 0.26 \pm 0.05$ corresponds to the vacancies being present in the sample just before melting, its value for the temperature interval 235–250 °C suggests that the defects acting as positron trapping centres in this temperature interval are much larger than monovacancies — most probably they are vacancy clusters or clusters of atoms conserving the crystalline structure. A decrease in the value of the R — parameter at 330 °C corresponds to the decomposition of these clusters.

4. Conclusions

The reconstruction of the structure of tin occurring during the melting process leads to the formation of large defects (most probably vacancy clusters or clusters of atoms conserving the crystalline structure). These clusters are stable in the temperature interval just above the melting point and decompose at about 400 °C, and are not formed anew during slow cooling of the sample soaked at 400 °C. The kinetics of this process is reflected by a sudden increase in the value of the R — parameter on passing through the melting point and further decrease with increasing temperature.

The vacancy formation energy determined in the present work equals 0.50 ± 0.02 eV.

This work has been supported by scientific program 01.06. CPBP.

References

Brandt, W., Waung, H. F.: Phys. Lett. **A 27**, 700 (1968)
Dedoussis, Sp., Charalambous, Stef., Chardalas, M.: Phys. Lett. **62A**, 359 (1977)
Herlach, D., Stoll, H., Metz, H., Jackman, T. E., Maier, K., Schaefer, H. E., Seeger, A.: Appl. Physics **12**, 59 (1977)
Kishimoto, Y., Tanigawa, S.: Positron Annihilation, Coleman, P. G., Sharma, S. C., Diana, L. N. (eds), Amsterdam 1982
Mantl, S., Triftshauser, W.: Phys. Rev. Lett. **34**, 1554 (1975)
Mascher, P., Puff, W.: Proc. CPA-7, New Delhi 1987
McKenzie, I. K., Lichtenberger, P. C.: Appl. Physics **9**, 331 (1976)
Rozenfeld, B., Chabik, St., Pajak, J., Jerie, K., Wierzchowski, W.: Acta Physica Polon. **A44**, 21 (1973)
Shah, N., Catz, A. L.: Phys. Rev. **B5**, 2498 (1984)
Szymański, Cz., Chabik, St., Pajak, J., Rozenfeld, B.: phys. stat. sol. (a) **60**, 375 (1980)
Szymański, Cz., Rudzińska, W., Chabik, St.: Zeszyty Naukowe WSP Fizyka XXII, 1986
West, R. N.: Adv. Phys. **22**, 263 (1973)

(Received April 23, 1987)

Author's addresses:

W. Rudzińska, Cz. Szymański, S. Chabik
Institute of Physics, Pedagogical University
Oleska 4
45052 Opole, Poland

B. Rozenfeld
Institute of Experimental Physics, Universiry of Wrocław
Wrocław, Poland

Cryst. Res. Technol.	22	1987	11	K185–K190

A. P. de Lima, C. Lopes Gil, D. R. Martins, N. Ayres de Campos, L. F. Menezes, J. V. Fernandes

Depto. de Fisica y Depto. Eng. Mecanica da Universidade, Coimbra, Portugal

Positron Studies in Polycrystalline Deformed Copper[1])

1. Introduction

Positron annihilation (PA) studies in deformed copper have been presented in numerous publications. Single crystals and polycrystalline samples with different purities have been studied mostly in terms of an average lifetime and Doppler broadening methods. The mobility of vacancies in electron irradiated samples have been identified to occur below room temperature (Mantl, Triftshauser). In well annealed samples a lifetime of about 122 ps has been consistently attributed to annihilations in the bulk. Dislocations have been associated with a lifetime around 170 ps. Longer lifetimes were also observed and attributed to the presence of larger defects. By resistivity measurements two recovery stages were observed in the temperature ranges from 250 to 400 °C and from 450 to 500 °C which were reasonably consistent with the evolution of the mean lifetime observed in positron annihilation measurements (Kopetskii et al.; Kupca et al.). However, the evidence for these stages seems to depend mostly on the initial treatments of the samples. An indication of these stages can indeed be seen in experiments using single crystal or polycrystalline samples and also for different purity of the samples (Myllyla et al.).

The aim of this work was to clarify the influence of an initial annealing on the evolution of the dislocation structures produced inside the grains by tensile deformation. The initial grain size after the previous annealing was measured by optical microscopy and the evolution of the defects was studied by PA and TEM.

2. Experimental

Samples were cut from a grade DHP commercial copper sheet 1 mm thick, 99.92% pure. After cleaning with acetone the samples were subjected to initial annealings for at least one hour and temperatures varying between 300 °C and 750 °C in a vacuum always better than 10^{-6} torr. The grain size was studied by optical microscopy after electropolishing with a solution of orthophosphoric acid (66% volume in water) at 15 V and surface etching at 2 V in order to reveal the grain boundaries.

Studies with positrons were also made with electropolished samples. The well annealed reference samples showed a single lifetime of 120.4(6) ps and a Doppler parameter $I_V = 0.20850(17)$.

Uniaxial tensile tests were performed up to 30% amount of strain on samples cut perpendicularly to the rolling direction of the commercial sheet for different previous annealings in the temperature range specified before. Grain size measurements made by optical microscopy were performed on these samples. Table 1 presents these results together with the corresponding I_V values from Doppler broadening measurements.

[1]) Contributed paper presented during the European Meeting on Positron Studies of Defects, March 23–27, 1987, Wernigerode, G.D.R.

On the basis of the results obtained in these preliminary studies it was decided to follow systematically, by positron annihilation and TEM, the evolution of the defects produced after 6.5, 15 and 27% degree of deformation and for the 400, 550, and 750 °C previous annealing temperatures.

Table 1
Grain size and I_v parameter after initial annealings

initial annealing temperature and time	300 °C 1 h	400 °C 1 h	500 °C 1 h	550 °C 1.25 h	600 °C 1 h	750 °C 1.50 h
grain size μm	20 ± 2	28 ± 2	34 ± 2	38 ± 4	75 ± 15	264 ± 20
I_v parameter	.2142(2)	.2116(2)		.2085(2)		.2085(2)

The deformed samples were cut in squares to 1 × 1 cm² and studied during isochronal annealing programs, starting at room temperature (25 °C) and in steps of 50 °C, till the complete recovery of defects was observed. Each annealing was performed during one hour in a vacuum better than 10^{-6} torr.

The samples for TEM observations were mechanically polished down to a thickness of 0.15 mm. Discs, 3 mm in diameter, were then cut and electropolished at 20 °C under 10 V till point perforation was reached. Microstructural observations were made using a 100 kV transmission electron microscope.

The samples for positron studies were electropolished for surface cleaning, and measured at room temperature after each annealing. A ^{22}Na positron source with ≅ 15 μCi activity, was used. Positron lifetime measurements were made in a fast timing system with 210 ps (FWHM) resolution and a count rate of about 100 cps. All spectra were taken with more than 2×10^6 counts.

Analysis of the time spectra was initially made with free parameters in order to obtain some general consistency of the results. A fixed source contribution with 8.4% intensity was considered. In most cases one lifetime varying between 260 and 280 ps with significant intensity was systematically obtained in the fits for the as-deformed samples and also for some sets of annealing steps. Some of these fits were considerably improved by adding another lifetime of the order of 1 ns. Since this long component was indeed not necessary for the analysis of the well annealed reference samples, the possibility of relating this component with surface effects had to be excluded.

The Doppler broadening system, based on an intrinsic germanium detector, had an energy resolution of 1.21 keV at the 497 keV peak from ^{103}Ru. The analysis of the Doppler spectra was made in terms of the parameter I_v integrated over a range of 3 mrad centered on the 511 keV annihilation peak.

3. Results and discussion

A summary of the I_v parameters obtained by Doppler analysis during isochronal annealing of the samples is presented in Figure 1. For samples with the same previous annealing the initial I_v value increases with the degree of deformation. Saturation of trapping was observed for the 27% degree of deformation. During the first two or three steps of annealing (below 125 °C), no significant change of the I_v parameter is observed. A continuous decrease of I_v is then present for the sample annealed at 400 °C (and also at 500 °C, not shown); however, for the sample annealed at 750 °C the I_v parameter stays during some annealing steps at a value higher than the obtained at room temperature. The complete recovery is obtained for consistently lower temperatures when the initial degree of deformation is increased.

TEM observations made on the deformed samples show that, even in the 6.5% deformed samples, the dislocations arrange themselves forming walls delimiting cells where the dislocation density is considerably small for all the observed grains. Most of the observed cells present a parallelepipedic shape as shown in Figure 2. However, some curved dislocation walls surrounding equiaxial cells were also seen. The cell size,

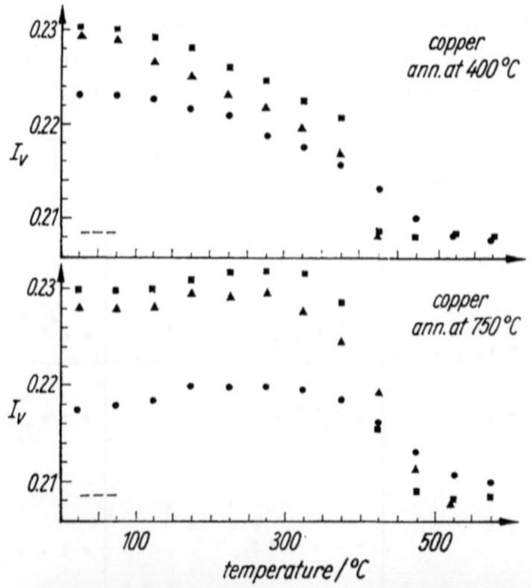

Fig. 1. I_v parameters during isochronal annealing for samples deformed to: ■ 27 %; ▲ 15 %; ● 6.5 %

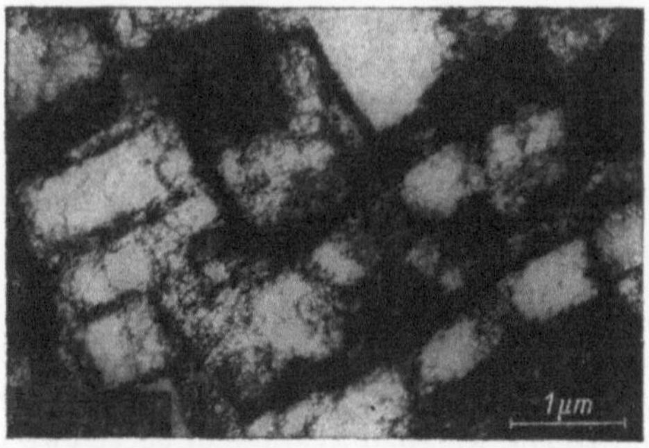

Fig. 2. Typical parallelepipedic cells as observed by TEM

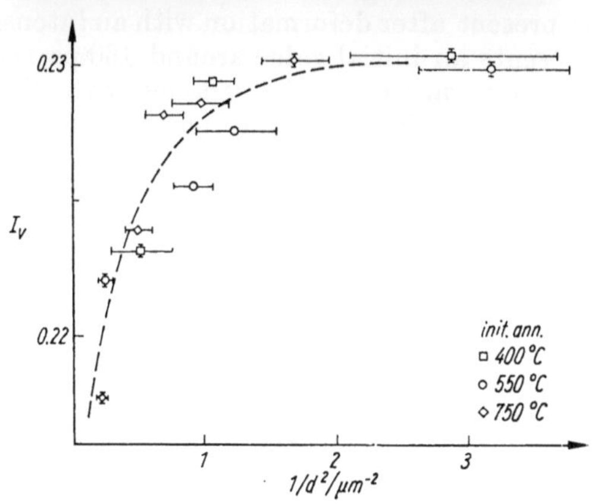

Fig. 3. I_v versus $1/d^2$ (d, cell diameter) for the deformed samples

determined as the minimum distance between two neighbour walls, is observed to decrease with the increase in strain. For the same amount of strain the cell size decreases when the initial grain size is smaller.

Experimental results and theoretical considerations (HOLT) have shown that the dislocation structure developed during the deformation tend to form cells the size of which, d, is proportional to the inverse square root of the dislocation density ϱ: $d = k\varrho^{-1/2}$. Since I_v is strongly dependent on the concentration of defects, we plotted in Figure 3 the evolution of I_v with the inverse square of the cell size which is a measure of the density of dislocations. Although I_v includes the effect of all the defects present the curve shows its dependence on $1/d^2$ before saturation effects are present.

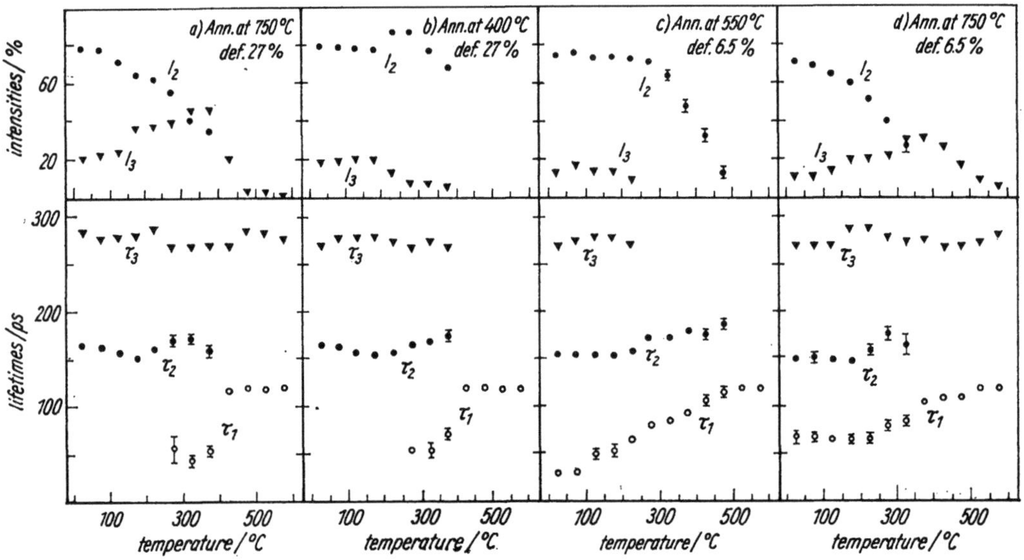

Fig. 4. Lifetimes and intensities as a function of the annealing temperature

Some examples of the lifetime results are presented in Figure 4. All the fits were made with a fixed lifetime of 270 ps with free intensity and, if necessary, introducing the 1 ns component. The intensity of this longer lifetime was always smaller than 0.6%. The τ_3 values presented correspond to the weighed average of these two long components. The lifetime τ_3 was always present after deformation with an intensity between 10 and 20%. The lifetime τ_2 presents an initial value around 160 ps showing some fluctuations with the annealing temperature. In the samples with 27% degree of deformation the lifetime τ_1 is not present for annealing temperatures below around 275 °C. This is a clear indication of saturation of positron trapping on the Doppler results (Fig. 1). The temperature for complete recovery of the defects is higher for the samples with 6.5% degree of deformation than for those deformed to 27%. Also the samples with similar deformation but a previous annealing at a higher temperature have a general tendency to present higher temperatures of recovery. The Doppler parameter results corroborate these observations.

These facts follow then the generally accepted tendency concerning interaction of defects such that, a higher density of defects decreases their effective migration energy which implies a lower temperature of recovery. In particular we mention that the sample with the smallest density of defects (6.5% degree of deformation after an initial annealing at 750 °C), does not show a complete recovery even at 575 °C (see Figs 4d and 1). It is important to refer here that TEM observations on this particular sample shows the recovery of dislocations but no recrystallization of the grains, as can be seen in Figure 5 where the existence of subgrain boundaries is evidenced. The recrystallization was observed by TEM, and confirmed by optical microscopy, for all the other observed samples.

The evolution of the defects associated with the lifetime τ_2 seems to be inconsistent with the general trend referred above. In fact, Figure 4 show that the decrease of I_2 in the sample deformed at 6.5% after an initial annealing at 750 °C, starts at a temperature lower than in the sample with the same deformation after the 550 °C initial annealing, where the initial density of defects is higher. Similar difference in the decrease of the intensity I_2 was observed in the samples with 27% degree of deformation where saturation of positron capture occurs. To the decrease of the intensity I_2 in the samples annealed at 750 °C corresponds a simultaneous increase of the intensity I_3. This seems to indicate that the mobility of the defects associated with τ_2, enhances the contribution of other defects with a characteristic lifetime of about 270 ps. These new defects will be dominant at about 325 °C and they control the final recovery in the samples with previous annealings at 750 °C.

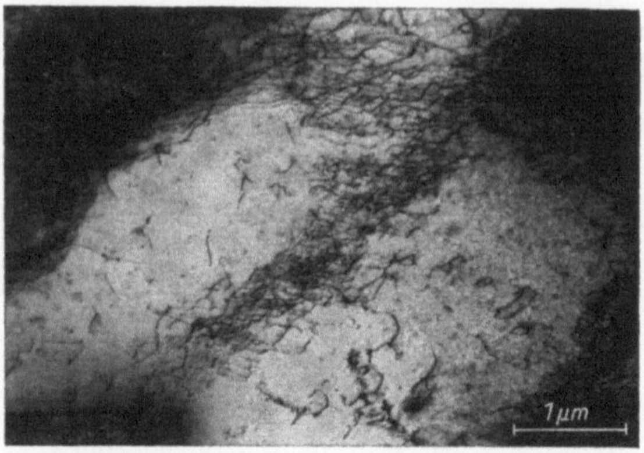

Fig. 5. Dislocation network typical of subgrain boundaries as observed by TEM

The identification of the defects associated with τ_3 is not clear. The value of τ_3 suggests the existence of defects with open volumes larger than single vacancies. They could be vacancy-clusters. It is well known that when dislocations move, during deformation or annealing, they can create vacancies which are mobile at room temperature. Vacancy-clusters could then be present after deformation. Since I_3 is nearly constant for annealing temperatures below 200 °C these clusters would be stable in this temperature region. As also observed by HINODE et al., some change occurs around 200 °C. In the samples with previous annealings at 400 °C and 500 °C, the τ_3 component starts to disappear which could be interpreted as a dissolution of the clusters. However, in the samples with previous annealing at 750 °C the component τ_3 remains till much higher temperatures. This can only be understood if the clusters are anchored at some type of precipitates which are only present after the 750 °C annealing.

The constancy of τ_3 and the increase of I_3 requires that the effective volume as seen by the positrons is mainly constant and that the trapping rate of the positrons increases. These arguments could lead us to include the dislocation loops as candidates for the interpretation of τ_3. However, the fact that this type of defects were not significantly observed by TEM certainly weakens these arguments. It is also difficult to explain this lifetime as due to some special arrangement of the dislocations, either including high density of jogs (DIAS et al.) or considering high stress fields in the region around them. The large value of τ_3 and its dependence on the previous annealing temperature are against this interpretation.

4. Conclusions

The values obtained for the lifetime of positrons annihilating in the bulk and in dislocations are in good agreement with previous results. The presence of defects with characteristic lifetime around 270 ps has been evidenced. This value, being much higher than the lifetime corresponding to dislocations, indicates the possible existence of larger defects. The enhancement of the influence of these defects in the samples with a previous annealing at 750 °C is still an open question. No direct evidence of the influence of grain boundaries and grain size of the positron annihilation parameters was seen. A correlation between the cell size and the Doppler parameter I_v was presented.

We would like to acknowledge for the facilities conceded on the use of the transmission electron microscope of the Faculty of Medicine of the University of Coimbra.

References

Dias, L., Pareja, R., Pedrosa, M. A., Prieto, J. I., Gonzalez, R.: Phil. Mag. **A51**, L61 (1985)

Hinode, K., Tanigawa, S., Doyama, M.: J. Phys. Soc. Jap. **42**, 1591 (1977)

Holt, D. L.: J. Appl. Physics **41**, 3197 (1970)

Kopetskii, Ch. V., Kulesko, G. I., Kokhanchik, L. S., Zharikov, O. V.: phys. stat. sol. **A22**, 185 (1974)

Kupca, S., Kerr, D., Hogg, B. G., Basinski, Z. S.: Can. J. Phys. **60**, 201 (1982)

Mantl, S., Triftshauser, W.: Phys. Rev. **B17**, 1645 (1978)

Myllyla, R., Karras, M., Miettinen, T.: Appl. Physics **13**, 387 (1977)

(Received April 23, 1987)

Authors adresses:

A. P. de Lima, C. Lopes Gil, D. R. Martins, N. Ayres de Campos
Depto. de Fisica da Universidade
3000 Coimbra
Portugal

L. F. Menezes, J. V. Fernandes
Depto. Eng. Mecânica da Universidade
3000 Coimbra
Portugal

Cryst. Res. Technol.	22	1987	11	K191–K194

Z. MICHNO, T. GORECKI

Physics Laboratory, Technical University of Opole, Opole

Positron Annihilation Studies of Binary Ni-Based Alloys[1])

Positron annihilation studies of Ni–1 at.% X (X = Pb, In and Zn) alloys have been performed. The experimentally observed changes in the parameters connecting the shape of angular correlation curve with the kind and concentration of crystal lattice defects are correlated with the values of the vacancy-solute binding energy calculated from the vacancy model of melting.

1. Introduction

The vacancy-impurity interaction is a very important factor affecting all the diffusion-controlled processes in alloys. However, our knowledge regarding the fundamental parameters describing the vacancy-impurity interaction is very meagre (GORECKI 1983).

One of the most active area of positron physics is their application to the studies of crystal lattice defects. Among others, the vacancy-impurity interaction in binary aluminium alloys was studied very intensively (SNEAD et al.; HALL et al.; KIM et al.; HASHIMOTO et al.). However, the extension of positron annihilation studies to a broad spectrum of various binary systems is required and various measurable positron annihilation parameters should be analysed using the same trapping model in order to establish the potentialities of the positron annihilation technique in studies of the vacancy-impurity interactions.

During the last decade the influence of many solutes on the behaviour of positron annihilation parameters during the recrystallisation, recovery and ageing of binary Ni-based alloys has extensively been studied by DLUBEK and coworkers (e. g. DLUBEK et al. 1973a, b, 1985; MEYENDORF et al.).

The purpose of the present paper is to report the results of our studies on the effect of some alloying additions (1 at.% of Pb, In, and Zn) on the shape of the angular correlation curves of positron annihilation quanta in binary Ni-based alloys. The results of measurements performed at high temperature, where the vacancy-impurity interaction should be clearly visible in the positron annihilation parameters, were compared with those taken at the room temperature, where the impurity induced changes in vacancy concentration are negligible, and the observed changes were correlated with the values of the vacancy-impurity binding energy calculated from the vacancy model of melting (GORECKI 1974, 1983).

2. Experimental

Nickel of the purity better than 4 N was used as basis material for producing binary alloys containing 1 at.% of Pb, In and Zn. The alloys were obtained by melting in induction

[1]) Contributed paper presented during the European Meeting on Positron Studies of Defects, March 23–27, 1987, Wernigerode, G.D.R.

furnace under inert atmosphere. The sample were prepared by cold-rolling to a thickness of 1.2–1.5 mm. Each of the samples was chemically etched and rinsed in distilled water and ethanol prior to the measurement. Part of the samples was subjected to 50 h tempering in vacuum (10^{-6} Torr) at 1273 K followed by slow cooling to the room temperature.

The angular correlation curves for the well-annealed samples of pure nickel and its alloys were measured at room temperature by means of a standard annihilation spectrometer with long-slit geometry of angular resolution of 1 mrad using one-side radiation of a ^{22}Na source.

The temperature dependence of the peak counting rate H was measured (with a resolution of 8 mrad) for cold-rolled samples, heated at a constant rate of 0.8 K/min under argon atmosphere up to 770 K. The measurements at the end temperature were continued until a stationary value of the peak counting rate was established. In the following this stationary value of H will be denoted by H_T. After reaching the thermal equilibrium the angular correlation curves were measured at 770 K. Note, that the angular correlation curves measured at 770 K were measured with angular resolution much worser as compared with those measured at room temperature.

3. Results and discussion

The experimentally determined angular correlation curves were deconvoluted into a sum of a narrow inverse-parabola, resulting from annihilation with conduction electrons, and a Gaussian corresponding to the annihilation with core electrons. The parameters describing the shape of the two components of angular correlation curve, measured at the room temperature are collected in Table 1, together with the parameters S and W as well as with the values of the vacancy-impurity binding energy

Table 1
Best-fit parameters of the angular correlation curves for well-annealed nickel and Ni–X alloys measured at room temperature

X	d (mrad) ± 0.08	θ_P (mrad) ± 0.15	I_N (a.u.) ± 0.003	S (a.u.) ± 0.003	W (a.u.) ± 0.002	B (eV)
Zn	6.15	5.59	0.197	0.560	0.094	0.07
In	6.22	5.44	0.179	0.551	0.098	0.16
Pb	6.13	5.72	0.199	0.560	0.106	0.26
Ni (pure)	6.36	5.66	0.198	0.558	0.111	—

calculated according to the procedure proposed by GORECKI. The S parameter is defined as the ratio of the area measured in the interval (−4, +4) mrad to the whole area under angular correlation curve, whereas the W-parameter is the ratio of the area measured in two symmetrically placed intervals (−12, −8) and (8, 12) mrad to the whole area under the angular correlation curve. θ_P and I_N denote the cut-off and the normalized intensity of the parabolic component and d is the standard deviation of Gaussian.

As it may be seen from Table 1, the introduction of 1% of solute changes the shape of the angular correlation curve, the effect of indium being the most prominent. In spite of the fact that the measurements were performed on the well-annealed samples at room temperature, the effect is hardly attributable to the changes in the concentration of defects acting as positron traps. The effect is the subject of further studies, the results of which will be published elsewhere.

Figure 1 represents the temperature dependences of the peak counting rate, normalized to the average value measured at 770 K, for cold-rolled Ni alloys. The measurements were performed at a constant heating rate of 0.8 K/min.

The shape of the curves registered for Ni–Zn and Ni–Pb alloys are very similar to those reported by DLUBEK et al. (1979a, b). However, our curve for Ni–In is more similar to that for Ni–Zn, whereas the curve for Ni–In reported by DLUBEK and coworkers is similar to that for Ni–Pb.

The S- and W-parameters of the angular correlation curves measured for the cold-rolled alloys at 770 K are represented in Figure 2 as a function of the value of the vacancy-solute interaction energy calculated according to the vacancy model of melting proposed by GORECKI (1974, 1983). As it is seen, the W-parameter decreases

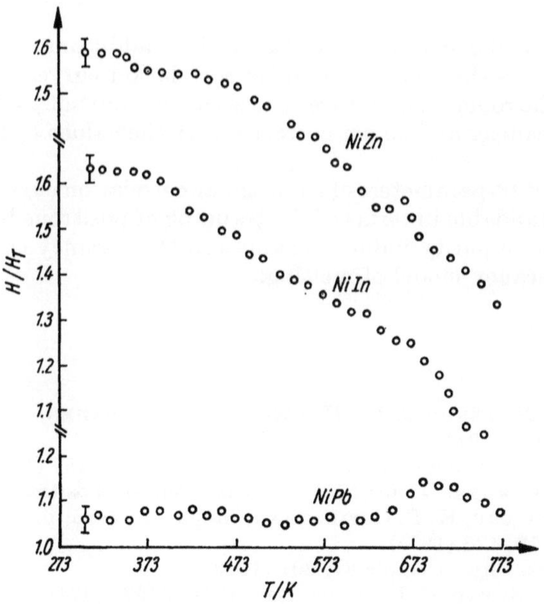

Fig. 1. Temperature dependence of the peak counting rate, for three different Ni–X alloys

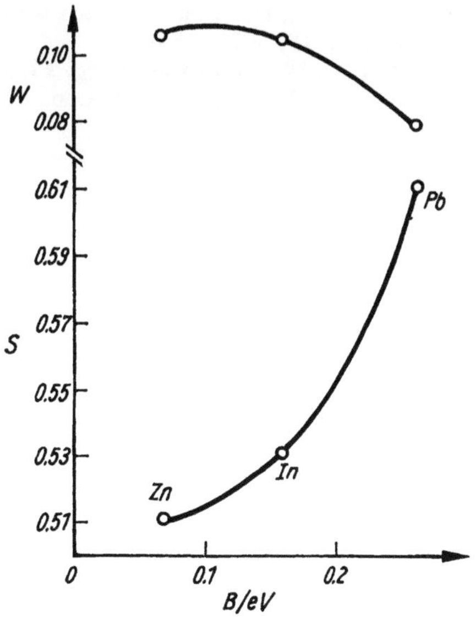

Fig. 2. The S- and W-parameters of the angular correlation curves, measured at 770 K versus the value of the binding energy of vacancy with solute atoms

and the S-parameter increases monotonically with increasing value of the vacancy-solute binding energy. As it is known, both the S- and W-parameters are sensitive to the vacancy concentration. The vacancy concentration in investigated alloys at 770 K, calculated according to the SCHAPINK's (1966) model, using our values of the vacancy-impurity binding energy is larger than that for pure nickel by a factor of 1.13, 1.62, and 7.65 for NiZn, NiIn, and InPb, respectively.

4. Conclusions

From the results presented above it follows that 1 at.% additions of Zn, In, and Pb in nickel markedly affects the shape of angular correlation curves measured on well-annealed samples at the room temperature. These changes are not connected with trapping of positrons in vacancies and do not correlate with the values of the vacancy solute-binding energy.

The changes in the S- and W-parameters of the angular correlation curves measured at 770 K are easily understandable in terms of the trapping of positrons by vacancies and correlate well with the impurity-induced changes in the vacancy concentration predicted basing on the vacancy model of melting.

References

DLUBEK, G., BRÜMMER, O., MEYENDORF, N., HAUTOJÄRVI, P., VEHANEN, A., YLI-KAUPPILA: J. Phys. F **9**, 1961 (1979a)

DLUBEK, G., BRÜMMER, O., MEYENDORF, N.: phys. stat. sol. (a) **53**, K157 (1979b)

DLUBEK, G., GORECKI, T., BRÜMMER, O.: in: Proc. 7th Int. Conf. on Positron Annihilation (eds P. C. JAIN, R. M. SINGRU, K. P. GOPINATHAN) Singapore 1985, p. 532

GORECKI, T.: Z. Metallkde. **73**, 426 (1974)

GORECKI, T.: Ber. Bunsenges. Phys. Chemie **87**, 801 (1983)

HALL, T. M., GOLAND, A. N., SNEAD, C. L.: Phys. Rev. B **10**, 3062 (1974)

HASHIMOTO, E., MORITA, Y., KINO, T.: J. Phys. F **14**, 879 (1984)

KIM, S. M., BUYERS, W. J. L., HOOD, G. M.: J. Phys. F **4**, 343 (1974)

MEYENDORF, N., DLUBEK, G., BRÜMMER, O., BARANOWSKI, A., ROZENFELD, B.: Cryst. Res. Technol. **19**, 507 (1984)

SCHAPINK, F. W.: Acta metallurg. **14**, 1131 (1966)

SNEAD, C. L., HALL, T. M., GOLAND, A. N.: Phys. Rev. Lett. **29**, 62 (1972)

(Received April 23, 1987)

Author's address

Doc. Dr. hab. Tadeusz GORECKI, Dr. Zbigniew MICHNO
Physics Laboratory, Technical University of Opole
45-370 Opole
ul. Ozimska 75
Poland

Cryst. Res. Technol.	22	1987	11	K195–K198

I. LEMAHIEU, L. DORIKENS-VANPRAET, M. DORIKENS, D. SEGERS

Laboratory for Nuclear Physics, Gent, Belgium

Defect Recovery after 77 K Plastic Deformation in Bismuth[1)]

Dopplerbroadening measurements were performed in 77 K deformed bismuth in the temperature range between 80 K and 410 K. An important trapping effect was observed. Two recovery stages were found, one at 230 K and the other at 380 K. The measurements were compared with the results obtained from positron annihilation lineshape measurements in as-received polycrystalline bismuth samples of different purities, and in room temperature deformed bismuth samples. In all samples a recovery stage at 380 K was found, which is ascribed to the annealing of dislocations. Lifetime measurements were performed in the as-received bismuth samples. For dislocations a lifetime of 289 ps was found.

1. Introduction

In recent years, a lot of defect studies have been performed in Bi, in order to investigate the behaviour of the point defects in this semimetal (see e.g. LEMAHIEU et al. and references therein). Due to the small overlap between conduction and valence bands, semimetals are characterized by the simultaneous presence of equal densities of electrons and holes. In bismuth this density is of the order of $3 \cdot 10^{23}\ m^{-3}$, which is five orders of magnitude smaller than in metals (KITTEL). As this density is so small, a tiny concentration of impurities can substantially alter the properties of bismuth. In spite of the many defect studies performed until now, a lot of questions still remain open, one of them being the location and the nature of the different recovery stages occurring in Bi. In this paper we report on positron annihilation lineshape and lifetime measurements performed both in 77 K deformed and room temperature deformed Bi and in as-received Bi samples of different purities.

2. Experimental

We have used Bi samples of different purities and different origins. All samples were polycrystalline material. The samples were made available from Koch & Light (Purity 6N), MRC (5N5), Highways (4N5) and Sidech (4N). Samples of 1.3 mm thickness were cut, cleaned in a solution of acetic acid, nitric acid and water (in the proportions 6:6:1) and then rinsed successively in diluted HCl, water and alcohol.

For the low temperature deformation, samples of purity 6N were annealed during 16 h in a vacuum of 10^{-4} Pa at a temperature of 410 K. The samples were deformed by compression under liquid nitrogen to a thickness reduction of 30%. After deformation a source (10 μCi of $^{22}NaCl$ on Al foil of thickness 0.005 mm) was mounted between two identical

---

[1)] Contributed paper presented during the European Meeting on Positron Studies of Defects, March 23–27, 1987, Wernigerode, G.D.R.

samples under liquid nitrogen and the source-samples assembly was put into a self-built gas-flow cyrostat, still under liquid nitrogen.

For the room temperature deformation, samples of purity 4N were used. The samples were deformed under air to a thickness reduction of 30% and afterwards mounted in the same cryostat as used for the low temperature deformation. For the measurement of the as-received Bi, the samples were also mounted in the same cryostat.

Dopplerbroadening measurements were performed on the deformed and the as-received Bi samples. The detector was an Ortec 81 cc hyper-pure Ge detector, and an ultra stable pulse generator was used for zero and gain stabilisation of the ADC (DORIKENS et al.). The resolution of the system was 1.22 keV for the 514 keV gamma line of ^{85}Sr at a count rate of $8 \cdot 10^3$ counts/s.

Lifetime measurements were performed on the 6N samples. The lifetime spectrometer is a fast coincidence system with a time resolution of 235 ps. As a source for this measurements we used 20 μCi of $^{22}NaCl$ evaporated on a 7.5 μm Kapton foil. The lifetime spectra were analyzed with the computer program PFPOSFIT (PUFF).

3. Results and discussion

a) Low temperature deformed Bi

The S-parameter characterizing the Dopplerbroadening of the annihilation line was measured in 77 K deformed Bi as a function of temperature between 80 K and 410 K, in steps of 10 K. Each measurement lasted one hour. Figure 1 shows the results of the measurements.

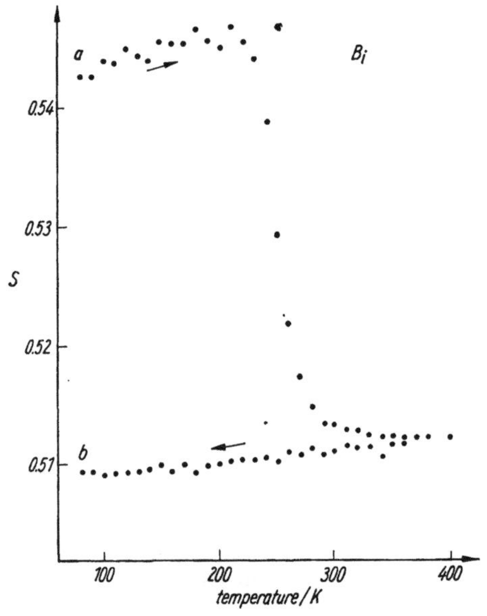

Fig. 1. S-parameters as a function of temperature in 77 K deformed Bi

Curve a was registered at increasing temperatures, and corresponds with the isochronal annealing of non-equilibrium defects introduced by the plastic deformation. The trapping effect observed here is very large. The results represented in curve b were registered at decreasing temperatures after the sample had reached for the first time a temperature of 400 K. Each point is a mean of two measurements. From this figure it follows that 2 recovery stages are present. The first is situated between 220 K and 290 K, and the activation enthalpy for this recovery stage was previously determined to be 0.50 eV (LEMAHIEU et al.). A second — much less pronounced — recovery stage is present between 370 K and 400 K.

In an electrical resistivity study of Bi quenched at liquid He temperatures, BITTAR and LESUEUR report two recovery stages, the first of them occurring between 150 K and 220 K, and the second occurring at 265 K. Electrical resistivity measurements were also performed in Bi irradiated with fast neutrons at 20 K (QUELARD, LESUEUR). Several recovery stages can be seen at lower temperatures (under 150 K), but no pronounced recovery stage was observed above 200 K. Recently CORBEL et al. have performed positron annihilation lifetime measurements in electron irradiated monocrystalline bismuth. The irradiation temperature was 20 K. The authors describe the presence of 2 stages in the recovery of the irradiated bismuth at annealing temperatures between 77 K and 400 K. The first stage, situated between 100 K and 125 K, is ascribed to the disappearance of vacancy clusters. A second stage occurs between 175 K and 200 K. Full recovery is only obtained at 350 K, and the authors state that at this temperature, all defects are completely eliminated.

We believe that the recovery stage found in our measurements between 220 K and 290 K corresponds with the stage at 265 K reported by BITTAR and LESUEUR. This stage should be ascribed to the annealing of vacancy-like defects. On the precise nature of this defects no statements can be made as yet.

The second recovery stage found in our measurements between 370 K and 400 K, was not reported previously. As the main defect structures introduced in a material by plastic deformation are dislocations, this stage can probably be ascribed to dislocations. This will be confirmed by the measurements on room temperature deformed and as-received bismuth.

b) Room temperature deformed Bi and as-received Bi

The Dopplerbroadening of the positron annihilation gamma ray was measured in bismuth samples as a function of temperature between 80 K and 520 K, in steps of 10 K. The samples were mounted at room temperature in the cryostat and then cooled down to 80 K to start the measurements. The samples were as-received polycrystalline material of purities 4N, 4N5, 5N5, and 6N. In all samples the same effect was observed, i.e. an annealing-out of a defect structure between 350 K and 400 K. Figure 2 shows a typical annealing curve. The dots represent measurements at increasing temperatures starting from 80 K up to 520 K. The circles represent measurements at decreasing temperatures after annealing at 520 K.

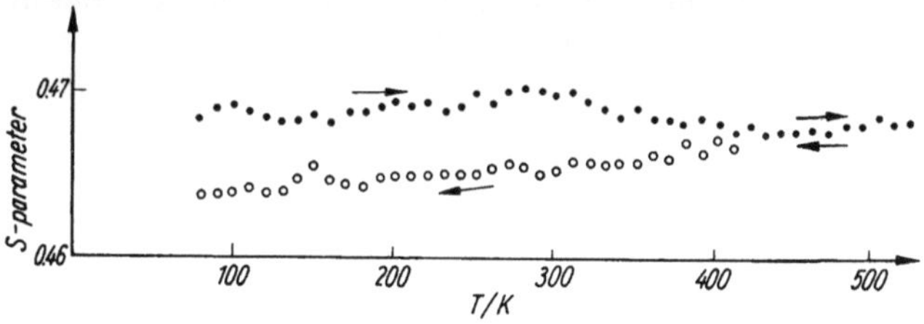

Fig. 2. S-parameter as a function of temperature in as-received Bi

The Dopplerbroadening of the positron annihilation lineshape was also measured in room temperature deformed Bi of purity 4N up to a thickness reduction of 30%. The deformation does not influence either the intensity of the effect or the annealing temperature. We believe that the stage found here is the same as the second recovery stage found in the 77 K deformed Bi. Positron annihilation lifetime measurements were performed on as-received Bi samples of purity 6N at room temperature. Those measurements were repeated after annealing of the Bi samples during 16 hours in a vacuum of 10^{-4} Pa at a temperature of 500 K. After annealing only one lifetime with a value of 234 ± 2 ps could be resolved. This lifetime corresponds well with the value of 239 ps obtained by CORBEL et al. in annealed Bi.

The lifetime spectrum of the as-received Bi samples prior to annealing could be resolved in 2 components. τ_1 has a value of 221 ± 13 ps and is ascribed to the annihilation of positrons in the bulk material. τ_2 has a value of 289 ± 36 ps and an intensity of 24.84%. From the trapping model (BERGERSEN, STOTT; CONNORS, WEST) it follows that, in the case that only one defect is present in the material, the trapping rate from the bulk to the defect is given by

$$k = (I_2/I_1)(\lambda_b - \lambda_2) , \quad (1)$$

where $I_1 = 1 - I_2$, λ_2 $(= 1/\tau_2)$ is the annihilation rate in the defect and λ_b is the bulk annihilation rate. Furthermore

$$\lambda_1 = k + \lambda_b , \quad (2)$$

where λ_1 $(= 1/\tau_1)$ is the observed annihilation rate of the shortest component. In order to check whether the two-state trapping model is applicable in this case, we will calculate λ_b^{calc} from eq. (1) and (2) and compare it with the experimental value λ_b^{exp} for the bulk annihilation rate. From eq. (1) and (2) it can be derived that

$$\lambda_b^{calc} = I_1\lambda_1 + I_2\lambda_2 , \quad (3)$$

and from the experimental values for the lifetimes τ_1 and τ_2 and the intensities I_1 and I_2 this gives us

$$\lambda_b^{calc} = 4.2604 \cdot 10^{-3}\ ps^{-1} \quad \text{or} \quad \tau_b^{calc} = 234.7\ ps .$$

As τ_b^{exp} has a value of 234 ps, it can be concluded that the two-state trapping model is applicable here, and that only one type of defects is present. We assume that these defects are dislocations, and they anneal out in Bi between 350 K and 400 K.

Further lifetime measurements on 77 K deformed Bi are in progress in order to obtain more information concerning the other recovery stages in Bi.

References

BERGERSEN, B., STOTT, M. J.: Sol. State Comm. **7**, 1203 (1969)

BITTAR, A., LESUEUR, D.: phys. stat. sol. (a) **48**, K123 (1978)

CONNORS, D. C., WEST, R. N.: Phys. Lett. **30A**, 24 (1969)

CORBEL, C., BOIS, P., MOSER, P., LEMAHIEU, I.: to be published in the Proc. of the 6th Int. Conf. on "Vavancies and Interstitials in Metals and Alloys", Berlin (West), eds H. WOLLENBERGER, C. ABROMEIT (Materials Science) 1986

DORIKENS, M., VAN DEN BOSSCHE, K., DORIKENS-VANPRAET, L., SEGERS, D., LEMAHIEU, I.: Proc. of the 7th Int. Conf. on Positron Annihilation, New Delhi, India, eds P. C. JAIN, R. M. SINGRU, K. P. GOPINATHAN, Singapore 1985, p. 819

KITTEL, C.: Introduction to Solid State Physics, 4th Ed. New York 1971

LEMAHIEU, I., DORIKENS, M., DORIKENS-VANPRAET, L., SEGERS, D.: Proc. of the 7th Int. Conf. on Positron Annihilation, New Delhi, India, eds P. C. JAIN, R. M. SINGRU, K. P. GOPINATHAN, Singapore 1985, p. 561

PUPP, W.: Comp. Phys. Comm. **30**, 359 (1983)

QUELARD, G., LESUEUR, D.: phys. stat. sol. (a) **36**, 729 (1976)

(Received April 23, 1987)

Authors' address:

I. LEMAHIEU, L. DORIKENS-VANPRAET, M. DORIKENS, D. SEGERS
Laboratory for Nuclear Physics
Proeftuinstraat 86
B-9000 Gent
Belgium

Cryst. Res. Technol.	22	1987	11	K199–K202

An. LIOLIOS, Sp. DEDOUSSIS, M. CHARDALAS, Stef. CHARALAMBOUS

Nuclear Physics and Elementary Particles Department
Aristotle University of Thessaloniki, Greece

Positron Annihilation in Thermally Treated Metallic Glasses[1])

The positron lifetime dependence on the annealing temperature has been studied for two transition metal-metalloid glasses (FeNi-based and Co-based). Small variations of the lifetime observed after annealing below crystallization temperature are attributed to elimination of the structural defects inherent in the as-quenched metallic glasses. More abrupt changes at higher temperatures are ascribed to crystallization processes which lead to polycrystalline phases. Differences in positron lifetime between annealed FeNi-based and Co-based metallic glasses are discussed.

1. Introduction

Transition metal-metalloid glasses are of great interest, especially for their technological applications as ferromagnetic materials (LUBORSKY, JOHNSON; WARLIMONT; EGAMI 1984). Thermal treatment changes significantly their magnetic and mechanical properties even far below crystallization temperature (WASEDA, EGAMI; PILLER, HAASEN). Specific knowledge of such changes is useful since in this way one can produce materials more suitable for special applications (HILZINGER, KUNZ), or prevent some undesirable changes of their properties (e.g. embrittlement (DENG, ARGON)).

Positron annihilation in metallic glasses shows an interesting behaviour which depends on the composition of the samples as well as on their thermal history (SHIOTANI; SCHILTZ et al.). In this work we present positron lifetime measurements as a function of annealing temperature for two metallic glasses different in their transition metal composition, i.e. an FeNi-based and a Co-based sample. The positron lifetime was measured after isochronal annealing at certain temperatures in the region 20 °C to 800 °C.

2. Experimental

Positron lifetime measurements were carried out using a fast-slow time spectrometer (resolution 240 ps). The positron source was about 15 μCi ^{22}NaCl sealed between two mylar foils (1 mg/cm^2). The source contribution was estimated (BISI et al.) for our samples to be about 10%, a fraction which fits satisfactorily to our experimental results. The lifetime spectra were analyzed by the program INTERACTIVE POSITRONFIT (VIRTUE et al.). Each spectrum had about 2×10^6 events.

The samples (1 × 1 cm) were cut from ribbons of two metallic glasses (FeNi-based: $Fe_{37}Ni_{37}Mo_4Si_6B_{12}$ and C-based: $Co_{67}Fe_4Mo_{1.5}Si_{16.5}B_{11}$). They are commercial products of

[1]) Contributed paper presented during the European Meeting on Positron Studies of Defects, March 23–27, 1987, Wernigerode, G.D.R.

VITROVAC, West Germany, prepared by a melt spinning technique. Fourteen layers stacked together on each side of the $^{22}NaCl$ source were required to make β-opaque samples. The positron lifetime spectra were collected at room temperature after isochronal annealing (30 min) at temperature T_a, in a vacuum about 10^{-6} Torr.

3. Results and discussion

We present our results as a function of annealing temperature T_a, in Figure 1. The lifetimes are yielded from a three-component analysis in which the two longer lifetime components (410 ps and 1800 ps) are fixed as source contribution.

Since the lifetime spectra have been collected at room temperature, any variation in the lifetime corresponds to irreversible structural changes induced by annealing at temperature T_a for 30 min. The lifetime shows a different behaviour for the two metallic glasses:

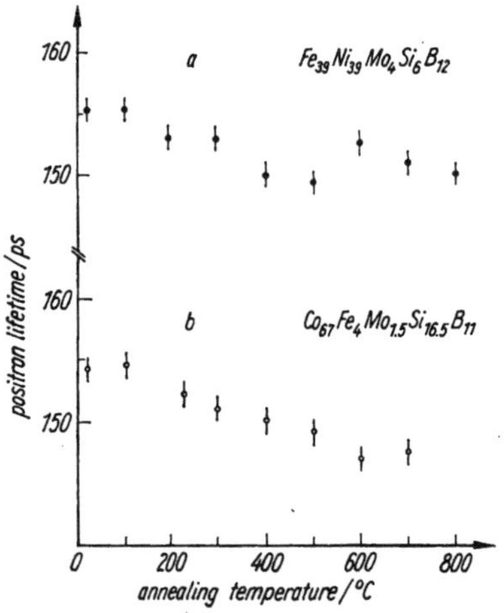

Fig. 1. The positron lifetime in $Fe_{37}Ni_{37}Mo_4Si_6B_{12}$ and $Co_{67}Fe_4Mo_{1.5}Si_{16.5}B_{11}$ as a function of isochronal annealing temperature

For the Co-based sample, curve (b) in Figure 1 shows a monotonous decrease which does not seem to change after annealing at crystallization temperature (500 °C).

For the FeNi-based sample on the contrary, curve (a) in Figure 1 shows a minimum near the crystallization temperature (450 °C) which is similar to the minimum observed for $Fe_{40}Ni_{35}Mo_4B_{15}$ (SCHILTZ et al.).

The decrease of the lifetime observed for both metallic glasses in the range between 100 °C and 300 °C, can be attributed to irreversible structural relaxation phenomena. Structural relaxation involves internal atomic rearrangements which leads to a more stable, although metastable, amorphous structure. EGAMI (1981) considers that these structural changes result from some quenched-in structural "defects" which starts to be annealed out during heating at temperatures even far bellow crystallization temperature. Such defects are local density fluctuations i.e. low density dilated regions ("n-type" defects) and high-density compressed regions ("p-type" defects) with continuous density distribution. A third type of quenched-in defects ("τ-type") are correlated with regions of high shear stresses. Reorientations of the local shear fluctuations cause only reversible changes, while changes of the local density fluctuations cause irreversible relaxation phenomena (JAGIELINSKI, EGAMI). Macroscopic irreversible changes, e.g. volume reduction upon annealing, for metallic glasses similar in composition with ours, was found to be of the order of 0.1% (ZEDLER, LEHMANN). Figure 1 shows that positron lifetime in both FeNi-based and Co-based samples decreases more than 1% after annealing at 200 °C. In order to explain this decrease we assume that positron lifetime in amorphous alloys is mainly determined by n-type defects. This assumption seems to be valid from the following arguments: n-type and p-type defects are not equally preferred as annihilation sites, since positrons are attracted by dilated regions and repelled by compressed ones. The mean annihilation probability depends on the characteristics of each type of defect, e.g. dilation and size, as well as their probability distribution. Amorphous state is full of such structural defects and positrons even at room temperature may visit many of them before annihilation. As positrons spend more time in dilated regions, most of them annihilate from sites in which the local density is relatively low. Annealing changes the topological short range order, so that a greater number of p-type and n-type defects have local density closer to the mean density value. If positron lifetime in amorphous alloys was determined equally from p- and n-type defects, annealing should leave the lifetime unchanged. The observed decrease of the lifetime could be explained if the role of n-type defects is more important (since the positron lifetime at these defects in higher).

Crystallization of FeNi-based amorphous alloys usually pass through some metastable intermediate phases before the final stable phases are obtained. Indeed, two stage crystallization is known to happen for $Fe_{40}Ni_{40}B_{20}$ (STUBICAR et al.), $Fe_{40}Ni_{35}Mo_4B_{15}$ (CUMBERA et al.) and $Fe_{32}Ni_{36}Cr_{14}P_{12}B_6$ (VON HEIMENDAHL, MAUSSNER). The first stage gives metastable phases, after annealing between 400 °C and 500 °C, with grain sizes above 200 nm (VON HEIMENDAHL, MAUSSNER; MÜLLER, VON HEIMENDAHL . It is reasonable to suggest that a similar first stage crystallization of $Fe_{37}Ni_{37}Mo_4Si_6B_{12}$, which takes place after annealing at temperatures between 400 °C and 500 °C, gives crystalline grains larger than the diffusion length of thermal positrons which, at room temperature, is about 100—200 nm (NIEMINEN). Based on the above presented hypothesis, the observed decrease of the positron lifetime is attributed to annihilation of a fraction of positrons in the bulk of these grains.

The second stage starts by recrystallisation at a higher temperature (between 500 °C and 600 °C) with the formation of grains of the final stable phases, usually at the boundaries of already existing grains. In order to explain the observed abrupt increase of the positron lifetime in this range, we suppose that the growth rate of the new crystallites is relatively small due to localized concentration of Mo atoms at the grain boundaries (SEAH). Annealing of the sample at 600 °C for 30 min is sufficient for the formation of only small crystallites, which means that positrons will be trapped and annihilate at grain boundaries. At temperatures above 600 °C, the formation of larger crystallites increases again the fraction of positrons annihilated in the bulk of the crystallites, and consequently gives the gradual decrease of the positron lifetime observed for annealing temperatures from 600 °C to 800 °C.

Crystallization of the Co-based sample will be interpretated according to results reported by WOLNY et al. for $Co_{70.3}Fe_{4.7}Si_{15}B_{10}$. They found that crystallization takes place through gradual formation of some phases such as a-Co, Co_2Si and Co_2B. An incipient crystallization of a-Co could be responsible for the gradual lowering of the positron lifetime at lower temperatures.

The lack of an abrupt increase of the lifetime for this metallic glass at a higher temperature is attributed to the direct formation of the final crystalline grains. In the opposite case of an intermediate crystallization stage, the decomposition of metastable phases and the formation of new ones would be responsible for the appearance of very fine grains, resulting in an increase of the positron lifetime. After annealing at temperatures above 500 °C, most of the positrons annihilate in the bulk of large crystallites.

4. Conclusions

Positron lifetime measured for a FeNi-based and a Co-based metallic glass is found to be affected more or less by the temperature of isochronal annealing. The slight decrease of the positron lifetime after structural relaxation has been considered as an indication that the role of n-type defects in positron annihilation is more important.

The positron lifetime behaviour after the beginning of crystallization depends on the size of the formated crystalline grains. For FeNi-based glass the abrupt increase observed at 600 °C is attributed to the formation of a new type of crystalline grains which have small size because of the segregation of Mo atoms at the boundaries of the previous grains. The monotonous decrease of the positron lifetime in the Co-based sample is ascribed to the gradual increase of the size of crystalline grains without passing through an intermediate crystallization stage.

References

BISI, A., GAMBARINI, G., ZAPPA, L.: Nuovo Cimento **53B** (2), 428 (1979)

CUMBERA, F. L., MIRANDA, H., CONDE, A., MARQUEZ, R., VIGIER, P.: J. Mater. Science **17**, 2677 (1982)

DENG, D., ARGON, A. S.: Acta Metallurg. **34**, 2011 (1986)

EGAMI, T.: Glassy Metals I. Ed.: H.-J. GUNTHERODT and H. BECK (Topics in Applied Physics, Vol. **46**) 1981, p. 25

EGAMI, T.: Rep. Prog. Phys. **47**, 1601 (1984)

HILZINGER, H. R., KUNZ, W.: J. Magn. Magn. Mater. **15–18** (III), 1357 (1980)

JAGIELINSKI, T., EGAMI, T.: J. Appl. Physics **55**, 1811 (1984)

LUBORSKY, F. E., JOHNSON, L. A.: J. Physique **41**, C8–820 (1980)

MÜLLER, K., VON HEIMENDAL, M.: J. Mater. Science **17**, 2525 (1982)

NIEMENEN, R. M.: Positron Solid State Physics. Ed.: W. BRANDT and A. DEPASQUIER (Proc. Int. Sch. Phys. "Enrico Fermi", Cource LXXXIII) 1983, p. 364

PILLER, J., HAASEN, P.: Acta Metallurg. **30**, 1 (1982)

SEAH, M. P.: J. Phys. F.: Metal Phys. **10**, 1043 (1980)

SCHILTZ, A., LIOLIOS, A., ELEFTERIADES, C., CHARDALAS, M., DEDOUSSIS, S., CHARALAMBOUS, S., CARTIER, E.: Proceedings 7th Int. Conf. on Positron Annihilation, N. Delhi. Ed.: P. C. JAIN, R. M. SINGRU and K. P. GOPINATHAN, 1985, p. 122

SHIOTANI, N.: Proceedings 6th Int. Conf. on Positron Annihilation, Texas. Ed.: P. G. COLEMAN, S. C. SHARMA and L. M. DIANA, 1982, p. 561

STUBICAR, M., BABIC, E., SUBASIC, D., PAVUNA, D., MAROHNIC, Z.: phys. stat. sol. (a) **44**, 339 (1977)

VIRTUE, C. J., DOUGLAS, R. J., MCKEE, B. T. A.: Comp. Phys. Commun. **15**, 97 (1978)

VON HEIMENDAHL, M., MAUSSNER, G.: J. Mater. Science **14**, 1239 (1979)

VON HEIMENDAHL, M., KUGLSTATTER, G.: J. Mater. Science **16**, 2405 (1981)

WARLIMONT, H.: Helvetica Physica Acta **56**, 281 (1983)

WASEDA, Y., EGAMI, T.: J. Mat. Science **14**, 1249 (1979)

WOLNY, J., ZAJAC, W., CALKA, A.: Nucl. Inst. Methods **199** (1, 2) 179 (1982)

ZEDLER, E., LEHMANN, G.: phys. stat. sol. (a) **81**, 445 (1984)

(Received April 23, 1987)

Authors' address:

An. LIOLIOS, S. DEDOUSSIS, M. CHARDALAS, S. CHARALAMBOUS
Nuclear Physics and Elementary Particles Department
Aristotelian University of Thessaloniki
54006 Thessaloniki
Greece

Cryst. Res. Technol.	22	1987	11	K203–K205

A. V. TSHERNYASHEVSKII, V. S. MIKHALENKOV

Academy of Sciences of the Ukrainian SSR
Institute of Metal Physics, Kiev, USSR

Positron Annihilation in Co Alloys Containing FCC and HCP Phases[1)]

Positron annihilation is studied in Co–Cu and Co–Mn alloys subjected to the heat treatments which result in the structural states with various lattices, phase contents and defects concentrations. The positrons are shown to be insensitive to the hcp layers in fcc lattice which form the stacking faults. Multiple transitions through the interval of martensitic transformation initiate the formation of defects of the higher concentration the larger of the volume effect of transformation.

Such defects of crystalline lattice as dislocations, vacancies and their agglomerates are known to trap positrons (HAUTOJÄRVI). Regarding the ability of stacking faults (s.f.) to trap positrons there are contradictory points of view (DEKHTYAR et al. 1965; MIKHALENKOV 1976; LYNN et al.).

The s.f. in fcc lattice represent the hcp layers restricted by partial dislocations. In order to ascertain the character of positron interaction with s.f. the contributions of partial dislocations and hcp layers to the total changes in annihilation parameters should be separated. Such separation may be attained by special processing of some alloys where a set of structural states, including multilayer ones, are formed during martensitic transformations. This concerns, in particular, Co alloys (NIKOLIN).

We studied angular correlation of annihilation photons (ACAP) in Co-6% Cu and Co-20% Mn alloys subjected to different treatments in order to create various structural states, e.g. fcc perfect lattice, fcc structure contained s.f. and the mixture of fcc and hcp lattices. The alloys have been melted in arc furnace using purified Ar atmosphere. The ingots were undergone forging and homogenizing annealing at 1370 K during 80 h. The 12 × 12 × 1 mm samples were cutted out of the forgings by electrical spark and then etched to remove disturbed layers. The Co–Cu alloy quenched from 1370 K into water has uniphase fcc structure which transforms into fcc + hcp one after cooling to 80 K. The multiple cycles between 670 and 80 K result in the multilayer trigonal symmetry lattice with large amount of random s.f.

Annealed Co–Mn alloy consists of both the residual fcc phase and martensitic one having hcp lattice. Unlike the Co–Cu, the martensite in Co–Mn alloy has considerably less random s.f. The structural states of the objects under studies were tested by X-ray analysis.

The ACAP curves were measured by conventional device with 1 mrad angular resolution and $3 \cdot 10^8$ Bq ^{22}Na positron source. The reduced area under parabolic fraction of ACAP $W = S_p/S_t$ was analysed together with r_m parameter which reflects the distance from lattice site at which positron and core electron wave functions product is maximum. This parameter is determined from the width of Gaussian frac-

[1)] Contributed paper presented during the European Meeting on Positron Studies of Defects, March 23–27, 1987, Wernigerode, G.D.R.

tion of ACAP δ by the relation (MIKHALENKOV 1974)

$$r_m = \sqrt{3/2}\,\frac{\hbar}{mc\delta},$$

where $\hbar$ is the Planck's constant, m is the electron mass and c is the velocity of light.

Such defects as vacancies and dislocations are known to cause the increase of both W (DEKHTYAR et al. 1964) and r_m (MIKHALENKOV 1974). If full dislocations are splitted into partial, and s.f. are formed between them a small growth of W is accompanied by decrease of r_m (MIKHALENKOV 1976).

The obtained results are listed in the Table 1.

Table 1
Positron annihilation parameters in Co–Cu and Co–Mn alloys subjected to different treatments

Alloy	Co – 6% Cu			Co – 20% Mn	
treatment	annealed at 670 K, slowly cooled	2 cycles between 670 and 80 K	20 cycles between 670 and 80 K	annealed	20 cycles between 670 and 80 K
structure	fcc perfect	fcc + hcp	fcc + s.f.	fcc + hcp	fcc + vac.
$W \cdot 10^3$	160 ± 5	180	180	150	210
r_m (pm)	71 ± 0.5	70	70	73	77

For annealed Co–Cu alloy $W = 0.16$ is very close to that ($W = 0.15$) estimated from W of pure Co and Cu using additivity rule. From this fact one can conclude that there are no trapping centers for positrons in this structural state. The W values for fcc + hcp and fcc + s.f. structures (Table 1, columns 2 and 3) are the same. Note that for the alloy under study the s.f. energy is so small that the width of s.f. is comparable with the grain size which was measured to be about 0.5 mm. Therefore s.f. involve the entire grains, and the last do not contain partial dislocations which move out to the grain boundaries. This point is supported by microscopic testing. Thus the effects of hcp layers and hcp admixture in fcc phase on positron annihilation parameters are the same. This can be considered as an argument that hcp layers themselves do not act as the traps for positron, and the effect of s.f. observed in papers (DEKHTYAR et al. 1965; MIKHALENKOV 1976) is connected with the positron trapping by partial dislocations. On the other hand, there is a distinction between positron annihilation parameters in perfect and hcp-contained fcc phase. We incline to attribute the small increase in W after thermal cycling between 670 and 80 K to the vacancies that are generated when the interval of martensitic transformation (below room temperature) is passed through. Although the volume effect of transformation in this alloy is small (a little bit higher than 0.1%) the large temperature interval of quenching (600 K) and repeated heating-cooling runs may promote the vacancies formation. Their concentration can be estimated on the basis of trapping model, according to which

$$\frac{W - W_0}{W_t - W_0} = \frac{\mu c}{\lambda_f - \mu c},$$

where λ_f and μc are the annihilation rate in perfect lattice and trapping rate, respectively, W_0 and W_t are the W values for free and trapped positrons.

Introducing some simplifications concerned W_t and μ values the vacancies concentration may be roughly estimated to be $c \cong 10^{17}\ \text{cm}^{-3}$.

For annealed Co–Mn alloy consisted of defectless fcc and hcp phases $W = 0.15$ is close to that for perfect fcc lattice in Co–Cu. However, the increase of W after thermal cycling is more pronounced reflecting more efficient accumulation of defects. Taking into account that the grains do not contain partial dislocations these defects are considered to be vacancies formed during multiple martensitic transformations. Applying trapping model their concentration is estimated to be $c \cong 6 \cdot 10^{17}\ \text{cm}^{-3}$, that is 6 times larger than in Co–Cu alloy. This corresponds to larger volume effect of transformation achieving 0.2%.

Above considerations are consistent with the behaviour of r_m parameter which increases when the positrons are trapped by defects. For fcc + hcp and fcc + s.f. structures r_m has the same values.

In conclusions the extended stacking faults in fcc lattice do not serve as the trapping centers for positrons. During multiple transitions through interval of martensitic transformation the more vacancies arise the larger is the volume effect of transformation.

References

DEKHTYAR, I. Ya., LEVINA, D. A., MIKHALENKOV, V. S.: Doklady AN SSSR (in Russian) **136**, 795 (1964)

DEKHTYAR, I. Ya., MIKHALENKOV, V. S., SAKHAROVA, S. G.: in: Fysitsheskaya priroda khrupkogo razrusheniya metallov (in Russian), Kiev 1965, p. 112

HAUTOJÄRVI, P. (ed.): Positrons in Solids, Berlin 1979 (Topics in Current Physics, vol. 12)

MIKHALENKOV, V. S.: Metallofysica (in Russian) **56**, 47 (1974)

MIKHALENKOV, V. S.: Metallofysica (in Russian) **64**, 3 (1976)

NIKOLIN, B. I.: Mnogoslojnyje structury i polytipism v metallitsheskych splavach (in Russian), Kiev 1984

LYNN, K. G., URE, R. W., BYRNE, J. G.: phys. stat. solidi (b) **60**, K117 (1973)

(Received April 23, 1987)

Authors' address:

Ing. A. V. TSHERNYASHEVSKII, Dr. sc. V. S. MIKHALENKOV
Institute of Metal Physics, Academy of Sciences of the Ukr. SSR
Str. Vernadskii, 36
252680 Kiev-142 USSR

Cryst. Res. Technol.	22	1987	11	K206–K209

T. J. PANEK, K. PLOTKOWSKI

Institute of Physics & Chemistry of Metals, Silesian University Katowice, Poland

Study of the Reversible Martensitic Transformation in Cu–Zn–Al Alloy by Positron Annihilation[1])

The positron lifetime measurement to investigate phase transformations: martensitic phase (MP) ⇔ parent phase (PP) in Cu–Zn–Al alloy has been used. We have observed the generation of new defects by MP → PP transformation; these defects are liquidated by martensitic transformation. This effect is only observed in case of high concentration vacancies in alloy. In order to explain the nature of these defects we take into consideration the supervacancies idea.

1. Introduction

Cu-14 at.% Zn–17 at.% Al alloy which after quenching from high-temperature β-phase reveals a martensitic structure, characterized with the appearance of the phenomenon of shape memory, has been the object of our investigation.

The temperatures of martensitic transformation and of the reversible one are, for the above composition, slightly above the room temperature. This alloy has already been investigated with techniques of positron annihilation. They enabled the estimation of influence of external strains upon the pseudoelastivity phenomenon (GEORGE et al.) and the changes of vacancy concentration in martensitic and parent phases (BOJARSKI et al. 1983) significantly influencing physical and mechanical properties of Cu–Zn–Al in the course of cyclic transformations MP ⇔ PP (MORAWIEC et al.).

2. Experimental

Before the measurements, samples have been annealed at 800 K and 850 K in protective argon atmosphere and then quenched to iced-water. The measurement of positron lifetimes has been done using a conventional fast-slow system of FWHM of prompt curve 310 ps. We have applied an external source of ^{22}Na positrons of about 0.3 MBq activity deposit on the sample. The stability of temperature ($\pm$0.3 deg) could be obtained during isochronal (80 ks) and isothermal measurements. The temperature of the laboratory has also been stabilized ($\pm$1 deg).

Slight but repeated changes of the values of annihilation parameters accompany the investigated process of phase transformations MP ⇔ PP have been observed (BOJARSKI et al. 1984). Thus much attention was paid not only to methods of measurements but also to high statistics of coincidency counts (more than 10^6 counts under the positron lifetimes curve was collected). Experimental spectra have been analysed numerically applying POSITRONFIT program fitting the lifetime component in decimal point of ps.

---

[1]) Contributed paper presented during the European Meeting on Positron Studies of Defects, March 23–27, 1987, Wernigerode, G.D.R.

3. Results and discussion

Figure 1A shows the results of isochronal measurements of positron lifetimes during heating (solid line) and cooling (dotted line) of Cu–Zn–Al in temperature range RT — 353 K. Below (Fig. 1B) we have shown the changes of intensity of longlife component. All experimental spectra, after substracted the positron source contribution, have been fitted into two components then mean lifetime of positrons $\tau = I_1\tau_1 + I_2\tau_2$ was calculated. The first component is due to positron annihilation in the vacancies of martensite phase $\{\tau_{1v}(\mathrm{MP}) \cong 183.2 \pm 0.8$ ps$\}$ or of parent phase $\{\tau_{1v}(\mathrm{PP}) = 179.5 \pm$ ± 0.8 ps$\}$. The other component is connected with the defects of lower electron density ($\tau_2 = 310 \pm 10$ ps). The course of average positron lifetimes reveals high (above the saturation density for positrons) concentration of vacancies frozen in martensite from a high-temperature β-phase. There have also appeared changes in τ connected with the martensitic transformation and the one reversible to it. Phase transformation are accompanied with the appearance or vanishing of the longlife component.

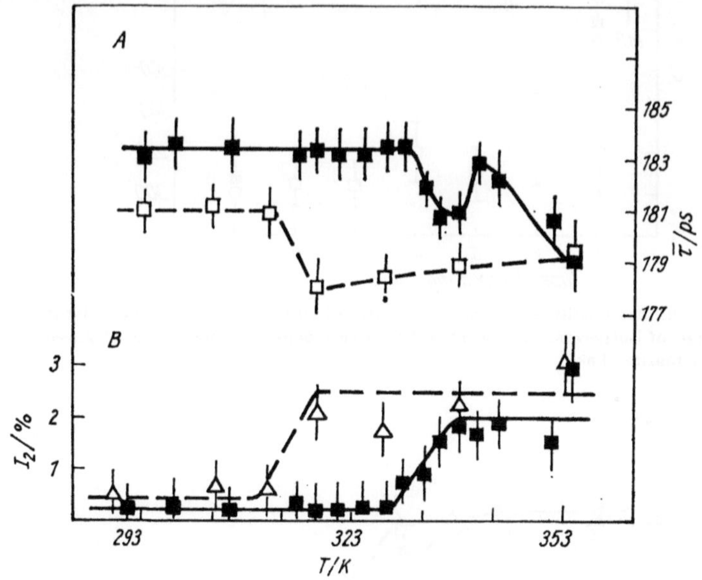

Fig. 1. The mean positron lifetime in Cu–Zn–Al (A) and intensity (I_2) of longlife component (B) as a function of heating temperature. Solid line — ascent temperature, dotted line — descent temperature

Positron lifetime in these defects is characteristic of large vacancy clusters. Such clusters could have been revealed by annealing of vacancies in the parent phase or could have been result of mobile vacancies clusterization. Still these two concepts are contradicted by the fact that the kinetics of the process is very fast and first of all the vanishing of longlife component during PP → MP transition. If the defects were vacancy clusters, their contribution to positron lifetime spectra during cooling should not visibly change.

In order to explain these doubts we have measured the positron lifetimes as the function of heating time at 356 K. For new samples of Cu–Zn–Al we have carried out testing measurements during heating from RT to 356 K. They showed MP → PP transformation in the analysed range of temperatures. The samples have been kept for two days at 356 K. This caused the initiation of clusterization process in Cu–Zn–Al.

After cooling the samples together with the furnace to RT, we have carried out two measurements of positron lifetimes. The intensity of longlife component equal about 2% is respective to frozen vacancy clusters obtained during the introductory measurement (black points in Figure 2). Going to measurements in the parent phase (after MP → PP) transformation leads to a rapid change of intensity of longlife component (to about 6%).

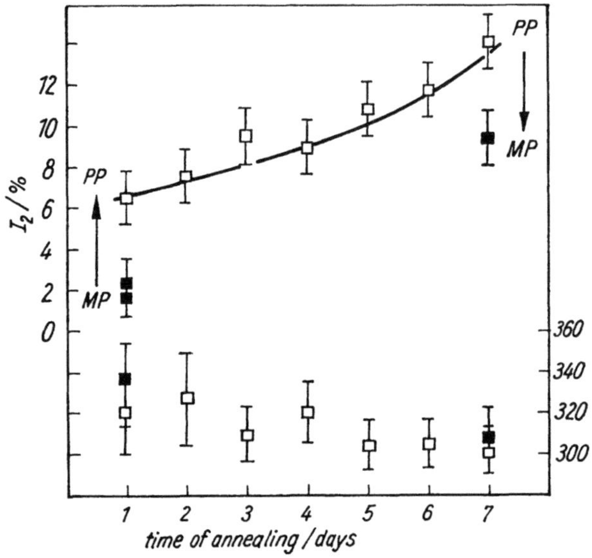

Fig. 2. The longlife component of positron lifetimes in Cu–Zn–Al and its intensity (I_2) versus time of isothermal heating at 356 K. Black points – the results of measurements at RT in martensite phase

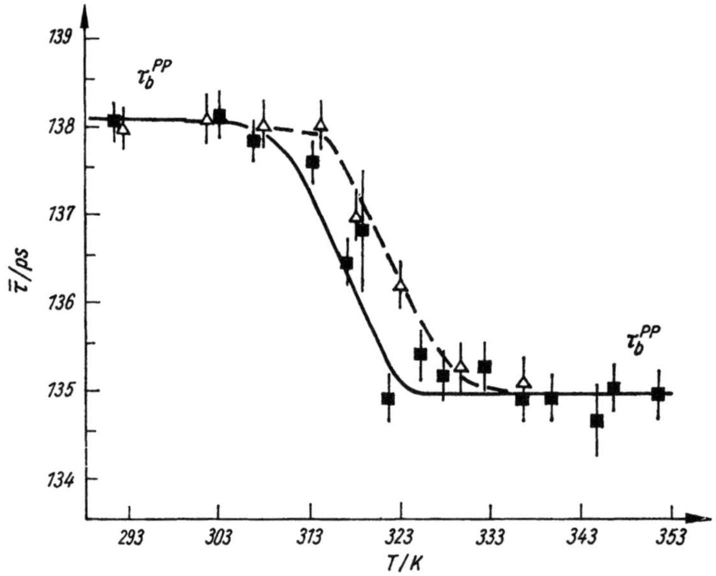

Fig. 3. The mean positron lifetime as a function of annealing temperature for sample after heating at 454 K in 120 s. Solid line – ascent temperature, dotted line – descent temperature

Annealing at 356 K shows the process of vacancy clusterization (Fig. 2A, circles). More than 14% of positrons annihilate in the clusters after 7 days of heating. Reversible transformation to the martensite phase (PP → MP) decreases the contribution of the longlife component about 4% which is respective to the liquidation of defects previously generated by MP → PP transformation. No significant changes of values of longlife component of positron lifetimes (Fig. 2) have been observed.

Further measurements proved the important function of vacancies in generating and in liquidating the defects with MP → PP transition. The Cu–Zn–Al sample has been heated at 454 K in 120 s. Then we have carried out a series of isochronal measurements of positron lifetimes in the time of heating function from RT to 356 K. The experimental data obtained from the whole range of temperatures showed the existence of a single time component connected with positron annihilation in free defects of martensite and parent phases. Thus the process of defects generating with the reversible martensitic transformation significantly depends on the high vacancy concentration.

Usually, the positron longlife component of 250 ps — 450 ps is connected with the large vacancy clusters in the examined material. Such an interpretation of τ_2 in case of Cu–Zn–Al cannot be accepted. It was suggested (Frank et al.) the possibility of existence of supervacancies with $\Omega > 1$ (Ω — atomic volume) in metals of b.c.c. structure in which the martensitic transformation appears. It seems that the formation and vanishing of extended and contracted local regions due to MP ⇔ PP transformations may cause the formation in the parent phase or disappearance in the martensitic phase of the supervacancies.

4. Conclusions

The large concentration of vacancies in martensitic phase of Cu–Zn–Al alloy cause appearing of some amount of supervacancies (sv) during martensite — parent phase transformation.

The vacancy — supervacancy transformation is the reversible process (v ⇔ sv) in copper base shape memory alloys.

This work was supported by the Institute of Experimental Physics of Warsaw University from CPBP 01.06 problem.

References

Bojarski, Z., Ilczuk, J., Panek, T., Morawiec, H.: J. de Phys. **44**, C9—241 (1983)

Bojarski, Z., Morawiec, H., Panek, T. J.: phys. stat. sol. (a) **83**, K97 (1984)

Frank, W., Seeger, A., Weller, M.: in: Positron Annihilation, eds P. G. Coleman, S. C. Sharma, L. M. Diana, Amsterdam 1982, p. 413

George, B., Hrynkiewicz, A., Janot, Chr.: Scripta metallurg. **17**, 723 (1983)

Morawiec, H., Bojarski, Z., Ilczuk, J., Panek, T., Augustyniak, M.: Archiwum Nauki o Materialach **5**, 31 (1984) (in Polish)

(Received April 23, 1987)

Authors' address:

Dr. T. J. Panek, Mr. K. Plotkowski
Institute of Physics & Chemistry of Metals
Silesian University Katowice
40-007 Katowice
Bankowa 12
Poland

Cryst. Res. Technol.	22	1987	11	K210–K213

I. LEMAHIEU, D. SEGERS, L. DESCHEPPER, L. DORIKENS-VANPRAET, M. DORIKENS, L. STALS, C. MOMMAERT, G. SEVERNE

Laboratory for Nuclear Physics, Gent, Belgium, and
Department of Materials Science, L.U.C., Diepenbeek, Belgium, and
V.U.B., Department of Physics, Brussel, Belgium

Annealing Study of He-Irradiated Ni Samples[1])

Positron annihilation lifetime measurements were performed in helium-irradiated Ni foils for annealing temperatures between 300 K and 1400 K. The samples were also examined by T.E.M. It is shown that immediately after the irradiations two kinds of vacancy clusters are present, some of them containing He-ions and the others being empty. The empty vacancy clusters anneal out before 800 K, while the vacancy clusters containing He grow in size as a function of annealing temperature.

1. Introduction

In recent times a lot of work has been performed on the study of radiation damage effects in stainless steels and metallic alloys. This research is of particular interest because of various material problems in reactor technology due to the presence of helium. Nickel has been widely used as a simple material that can serve as a model for these studies (see e.g. NAIDU et al., and references therein). As a lot of mechanical and other properties of irradiated metals and alloys are in many cases determined by the point defect controlled diffusion processes, it is clear that direct information on the annealing behaviour of defects in metals and alloys, together with the influence of ion impurities on this behaviour, can be quite useful.

The positron annihilation technique has been applied successfully to the investigation of the recovery behaviour of defects in metals and alloys. In this paper we study the defect recovery in alpha-irradiated nickel specimens by the positron annihilation lifetime technique.

2. Experimental

Ni foils of a thickness of 50 μm were irradiated with high energy He ions at the cyclotron facility of the V.U.B.-Brussels (Belgium). An energy degrader was used to obtain helium ions with a variable energy below 20 MeV. The irradiation was done up to a dose of about 600 ppm ($2.7 \cdot 10^{17}$ He/cm^2). The purity of the sample material was 5N.

Positron annihilation lifetime measurements were performed immediately after the irradiation and after annealing at temperatures ranging from 400 K to 1400 K in intervals of 100 K. The lifetime spectrometer is a fast coincidence system with a time resolution of 235 ps. A source (20 μCi of ^{22}NaCl on a Kapton-foil of thickness 7.5 μm) was mounted between two irradiated foils. This assembly was sandwiched between thicker nickel samples ($10 \times 10 \times 0.25$ mm) in order to stop all the positrons. The annealings were performed in a vacuum of 10^{-4} Pa. The annealing time was half an hour, and after each

[1]) Contributed paper presented during the European Meeting on Positron Studies of Defects, March 23–27, 1987, Wernigerode, G.D.R.

annealing the samples were cooled down to room temperature to perform the measurements. The lifetime spectra were analysed with the computerprogram PFPOSFIT (PUFF).

3. Results and discussion

Immediately after the irradiation the lifetime spectrum (see Fig. 1) contained a first lifetime component of 174 ps and a second of 330 ps, with an intensity of 31%. Between 400 K and 800 K the second lifetime decreases, and so does its intensity. Above 800 K the second lifetime component increases from 260 ps to 580 ps. Its intensity jumps from 22% to 53% between 700 K and 800 K, and decreases continually to 4% between 800 K and 1400 K. The first lifetime τ_1 decreases from 174 ps, immediately after the irradiation, to about 110 ps at 1400 K.

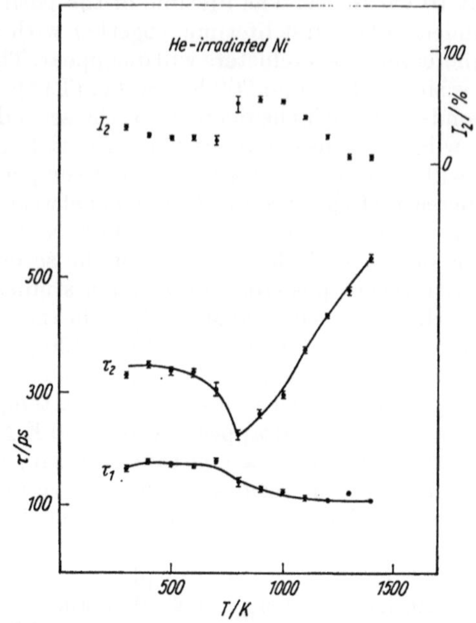

Fig. 1. Positron annihilation lifetime parameters as a function of annealing temperature in He-irradiated Ni

As the free lifetime in pure Ni has a value of 110 ps (NAIDU et al.; DLUBEK et al.), it is remarkable that the first lifetime never reaches a value beneath 110 ps. From the trapping model it follows that the first lifetime τ_1, corresponding to the annihilation of positrons in the bulk material, should have a value that is lower than the free lifetime. This is not the case here, which is an indication that either the defect concentration is so high, that all positrons are trapped in defects before annihilation, and there is no annihilation from the bulk, or the first lifetime τ_1 in our result is the mean of at least 2 components that can't be separated in our analysing computer program.

As the second lifetime has a value of 330 ps immediately after the irradiation, this lifetime should be ascribed to vacancy-like clusters, and we immediately think of HeV-clusters, i.e. vacancy clusters which contain one or more He impurities. In that case, however, we can't explain the decrease of this second lifetime τ_2, as we expect that HeV-clusters will grow with increasing annealing temperature. And what's more: in T.E.M. pictures taken at the L.U.C. — Diepenbeek on the same samples, no bubbles are visible in the pictures taken at 300 K. The first bubbles only become visible after annealing at 800 K. Thus we are faced with a contradiction. On the one hand we know that the lifetime increases as the size of the clusters increases, but on the other hand we don't see any bubbles in the T.E.M. pictures taken at 300 K, where the lifetime is long. But we do see bubbles in the T.E.M. pictures at 800 K, where the lifetime is short, and as a consequence the bubbles should be small. So it is necessary to introduce an additional effect into the model. Therefore we suppose that by irradiation, not only HeV-clusters were created, but also empty vacancy clusters, called V-clusters. As the lifetime corresponding with the annihilation of positrons in V-clusters is always higher than the lifetime in HeV-clusters of the same size (JENA, PONNAMBALAM), the second lifetime τ_2 should be associated with the annihilation in V-clusters. This is in accordance with theoretical calculations by Puska and Nieminen (PUSKA, NIEMINEN), from which we can derive that the V-clusters in our as-irradiated specimens should consist of about 10 vacancies. The lifetime of the positrons in the HeV-clusters should then be included in the first lifetime, together with some other components. As the temperature increases, the V-clusters will disappear. The temperature where voids in Ni disappear is situated around 700 K—800 K (TANIGAWA et al.; BEASLEY et al.), and this corresponds well with the decrease of the second lifetime τ_2 in our results. The HeV-clusters will not anneal out, as they are stabilized by the presence of the He gas, and they will grow in size as a function of temperature. This effect is clearly observed in the increase of the second lifetime τ_2 between 800 K and 1400 K. As the HeV-bubbles are growing between 800 K and 1400 K, their density decreases, which is reflected in the decrease of the intensity of the second lifetime. We have compared our findings with recent positron annihilation studies on alpha-irradiated nickel by NAIDU et al., who had used a 30 MeV alpha beam and a much lower dose of $0.7 \cdot 10^{17}$ He/cm^2. After irradiation they find a value of 420 ps for τ_2 which they ascribe to trapping in microvoids. The recovery of these microvoids was observed to lie between 400 K and 650 K, which is somewhat lower than the temperature at which the vacancy clusters in our measurements disappear. Above 650 K NAIDU et al. find that τ_2 lays between 340 ps and 400 ps, with a constant intensity of approximately 15%. This is not in agreement with our results, as we observe in the higher temperature region an increase in the second lifetime up to 580 ps and a decrease in its intensity.

We have also compared our results with earlier work on alpha irradiated nickel by VISWANATHAN et al. Their alpha beam had an energy of 28 MeV and the implanted helium dose was $4 \cdot 10^{17}$ He/cm^2 which resulted in a concentration of 500 at. ppm, which is very comparable with our value of 600 at. ppm. Up to an annealing temperature of 1000 K, their results are qualitatively the same as ours. They give, however, a different explanation for the trapping of positrons in the second lifetime immediately after the irradiation, as they rule out assignment of τ_2 to a vacancy cluster. This was based on the fact that τ_2 is observed at a temperature somewhat lower than the onset of monovacancy migration in nickel. We think, however, that vacancy clusters may be formed by displacement cascades during irradiation. At higher temperatures (above 1000 K) VISWANATHAN et al. obtain three lifetimes. The second lifetime increases and saturates at a value of approximately 300 ps. This saturation lifetime was ascribed to positron surface states in the formed He bubbles. A third lifetime, with a value between 600 ps and 800 ps, was attributed to the pick-off annihilation rate of ortho-positronium formed in large helium bubbles. In our measurements no more than 2 lifetimes could be resolved.

4. T.E.M. results

Identical Ni foils were irradiated under the same conditions to be examined by Transmission Electron Microscopy at the L.U.C.-Diepenbeek. Their results seem to confirm our conclusions, and give some additional information. T.E.M. pictures were taken after irradiation at 300 K, and after annealing until 700 K, 800 K, 900 K, 1000 K, and 1100 K. In the first picture (300 K), a very high density ($\sim 10^{23}/m^3$) of small loops is present, most probably of interstitial type, with a diameter less than 100 Å. There are also some small defects with triangular strain fields present and their diameter is not greater than 60 Å. It is not clear whether these are interstitial type defects or not. The lifetime corresponding with the annihilation of positrons in the loops, should be included in our first lifetime. This first lifetime should thus be a mean of the lifetime in HeV clusters, the lifetime in the loops, and possibly the lifetime in the bulk material. The V-clusters, associated with the second lifetime, can't be seen by the T.E.M. technique, as their size is too small.

In the pictures taken at 700 K and 800 K, one can see a decrease in the density of the loops, and an increase in the loop diameter. At 900 K, there are still loops present, but one can see that on the one hand various dislocation lines are growing, and that on the other hand some small bubbles were created. These bubbles indicate the start of the nucleation process. In the last pictures, taken at 1000 K and 1100 K, there are many dislocation lines, and a high bubble density ($\sim 10^{22}/m^3$) can be observed along the dislocation lines. It's remarkable that even at such high temperatures, dislocations do not anneal out. These dislocations, however, aren't observed in the positron lifetime measurements.

Annealing experiments on Ni samples irradiated to about 100 ppm and to about 3000 ppm are in progress.

References

BEASLEY, P., ALAM, A., USMAR, S. G., COLEMAN, P. G., WEST, R. N.: Proc. of the seventh International Conference on Positron Annihilation, New Delhi, India, (1985), editors: P. C. JAIN, R. M. SINGRU, K. P. GOPINATHAN, (World Scientific, Singapore), p. 114

DLUBEK, G., BRÜMMER, O., MEYENDORF, N., HAUTOJÄRVI, P., VEHANEN, A., YLI-KAUPPILA, J.: J. Phys. **F 9**, 1961 (1979)

JENA, P., PONNAMBALAM, M. J.: Phys. Rev. **B 26**, 5264 (1982)

NAIDU, S. V., SEN GUPTA, A., MUKHOPADHYAY, G., BHANDARI, R. K., SEN, P.: Rad. Eff. **83**, 129 (1984)

PUFF, W.: Appl. Phys. **18**, 165 (1979)

PUSKA, M. J., NIEMINEN, R. M.: J. Phys. **P 13**, 333 (1983)

TANIGAWA, S., SHINTA, I., IRIYAMA, H.: Proc. of the Yamada Conference V on Point Defects and Defect Interactions in Metal, Kyoto, Japan, (1982), editors J. TAKAMURA, M. DOYAMA, K. KIRITANI, Amsterdam, p. 736

VISWANATHAN, B., TRIFTSHÄUSER, W., KOEGEL, G.: Rad. Eff. **78**, 231 (1983)

(Received April 23, 1987)

Authors' addresses:

I. LEMAHIEU, D. SEEGERS, L. DORIKENS-VANPRAET, M. DORIKENS
Laboratory for Nuclear Physics
Proeftuinstraat 86
B-9000 Gent, Belgium

L. DESCHEPPER, L. STALS
Department of Materials Science, L.U.C., Universitaire Campus
B-3610 Diepenbeek, Belgium

C. MOMMAERT, G. SEVERNE
V.U.B., Department of Physics
Pleinlaan 2
B-1050 Brussel, Belgium

Cryst. Res. Technol.	22	1987	11	K214–K218

G. SOLT, W. B. WAEBER, A. SCHENCK, F. N. GYGAX, S. BARTH, F. FRISIUS

Swiss Federal Institute for Reactor Research, Würenlingen, Switzerland, and
Institute for Medium Energy Research, ETH, Villingen, Switzerland, and
Research Centre, Geesthacht, FRG

Observation of Precipitates in a Supersaturated Fe–Co Alloy by Muon Spin Rotation Spectroscopy: a Step towards Understanding Embrittlement of Reactor Pressure Vessel Steels[1]

The technique of μ^+SR spectroscopy was applied to detect precipitate formation due to ageing in the binary Fe-0.8 wt% Cu alloy. As compared to the homogenized specimen, the aged sample gives a μ^+SR signal with considerably larger relaxation rate in the temperature interval 180 K $< T <$ 350 K, having a maximum near $T = 250$ K with $\Lambda = 14\ \mu s^{-1}$.

Since precipitation of copper plays possibly an important role in the embrittlement of neutron irradiated ferritic steels, the signal from an irradiated steel specimen has also been investigated.

1. Introduction

Bombarding the material by energetic particles (like fast neutrons with energy $E > 1$ MeV) may lead to the formation of defects (voids, clusters) which are stable up to relatively high temperatures, say 400 °C in our example. The presence of these defects is revealed by a change in macroscopic properties, normally by hardening of the metal. One of the most intensively studied case, because of its technical importance, is that of the irradiation hardening and embrittlement of copper containing (0.1 to 0.3 wt %) ferritic steels (STEELE), used also in reactor industry. The intriguing problem here is that, although copper content is found to be intimately correlated with the observed embrittlement, no precipitates or any other microstructural changes following irradiation could be seen by transmission electron microscopy, and the very small sized ($R = 10-20$ Å) irradiation defects observed by small angle neutron scattering (FRISIUS, NARAGHI; BEAVEN et al.; SCHWAHN, PACHUR; LITTLE; LUCAS, ODETTE; SOLT et al.) are found not to be simply copper precipitates but particles of a more complex structure (like copper-vacancy agglomerates, for example).

Due to the difficulty in predicting the response to irradiation of a multicomponent system, it seems plausible to investigate binary Fe–Cu alloys first. Previous results on binary model alloys (FRISIUS et al.) for the relevant copper concentration $c = 0-1.0$ wt%, solutionized initially by a heat treatment above 800 °C, can be summarized as follows:

---

[1]) Contributed paper presented during the European Meeting on Positron Studies of Defects, March 23–27, 1987, Wernigerode, G.D.R.

i) below $c = 0.29$ at.% the defects generated by fast neutron irradiation are voids, like in pure iron, with radius $R \sim 6$ Å;
ii) copper precipitates, with mean radius 20–26 Å, appear in the irradiated alloys for $c > 0.29$ at.%. The radii increase, within this interval, with copper content;
iii) thermal ageing (e.g. 200 h/500 °C) of the homogenized samples causes precipitation only for a relatively high ($c > 0.64$ at.%) copper content. The particles are now much larger, the radius growing with c from 50 Å up to ~ 90 Å at $c = 0.86$ at.%. The precipitated volume fraction is always below the nominal copper content, of about 20%.
iv) subsequent irradiation of the pre-aged samples generates new, additional precipitations, with radii in the range of 15–25 Å, i.e. similar to those in the samples with no ageing;
v) in either case, after the irradiation practically no more copper is left in solution.

Pure copper particles in iron solvent are easily identified by small angle neutron scattering (SANS) via the ratio of their nuclear and "magnetic" contrasts – for more complex defects, however, the SANS pattern as taken alone may be consistent with several interpretations concerning the defect composition, and this is indeed the case for the results on steel (LUCAS, ODETTE; SOLT et al.). Hence, complementary information is needed: positron annihilation (HIGHTON; GHAZI-WAKILI et al.) has, for example, already been used with success in observing radiation damage in neutron irradiated metals and also in steel, in particular by detecting the evolution of the signal with subsequent annealing.

The recent success of μ^+SR spectroscopy in studying radiation damage in metals (MOESLANG et al.; WEIDINGER) has motivated us to use this method also for the study of the embrittled state of ferritic alloys. As a first step, we look for the muon signal at samples with thermally induced copper precipitates, in view of a further comparison with neutron irradiated specimens. For very high purity alloys one expects separate frequencies for each kind of defects (MOESLANG et al.), for less pure samples only the change in relaxation rate due to the given kind of defect becomes observable, as seen below.

2. Experimental

The method μSR, with emphasis on solid state physics application, has been described in detail (SCHENCK). The present experiments were performed at the 600 MeV proton facility of the SIN. A polarized muon beam was stopped in the Fe–Cu and steel samples at zero external magnetic field, and the evolution of polarization with time was detected by two detectors counting the positrons, emitted at each muon decay event, in the forward (f) and backward (b) directions.

Due to the magnetization in each domain the muon spins precess, giving rise to an oscillating intensity at a given direction, and the analysis of data was performed by assuming that the number N of the counts varies with time as

$$\frac{dN_{f/b}}{dt} = \frac{N_0}{\tau} e^{-t/\tau} \left[1 + AP(t) \cos(\omega_\mu t + \varphi)\right],$$

where $\tau = 2.2$ μs is the muon lifetime, A is the asymmetry parameter, $\omega_\mu = 2\pi \cdot (13.55$ MHz/kG$) \cdot B_\mu$ is the precession frequency determined by the (average) local magnetic field B_μ at the muon position and $P(t)$ is the dephasing function assumed to have the form $P(t) = \exp(-\Lambda t)$. Here Λ is the rate of relaxation, characteristic of trapping and eventual detrapping of the muon at defect sites. After removing the trivial exponential factor, the experimental intensity-time curves are fitted in the standard way to find A, ω_μ, Λ, and φ; the spectral density $I(\omega)$ is also determined.

3. The samples

In the first experiments the following samples were studied:

a) polycrystalline "pure" (99.9%) iron, the ground material for Fe–Cu alloys;
b) homogenized Fe-0.8 wt% Cu alloy with the copper in solution. The sample was solutionized by a heat treatment at 840 °C/48 h and quenched to prevent precipitation;
c) the same alloy as b) but subjected to an ageing treatment at 500 °C/120 h so as to bring about precipitation of the copper. According to previous SANS results (Frisius et al.), the precipitated copper particles in this case have mean radii of 60 – 70 Å, the size distribution has a half width of about 20 Å;
d) reactor steel (main alloying elements in wt%: C: 0.172, Si: 0.29, Mn: 1.47, Ni: 0.87, Mo: 0.54, Cu: 0.173); one of the two identical specimens of this material was previously irradiated in a research reactor by fast ($E > 1$ MeV) neutrons at a dose $\Phi = 1.9 \times 10^{19}$ n/cm².

The temperature of the samples ($12 \times 16 \times 1.5$ mm³, contained in an Al holder) was controlled from 180 K to 350 K.

4. Results

As to the precession frequencies, we have obtained ω_μ (Fe; $T = 300$ K) $= 48.65 \pm 0.05$ MHz at room temperature for the iron sample: this compares well with previous data of 48.66 and 48.44 MHz (Stronach et al.; Graf et al.).

The frequencies for both the homogeneous and precipitate-containing alloy phases are lower, by $\Delta\omega \sim 1$ MHz, and increase slowly with decreasing temperature, following the trend of $\omega_\mu(T)$ for pure iron — we interpret $\Delta\omega$ as a simple "dilution" shift.

The relaxation rates, plotted in Figure 1, show the following main features:

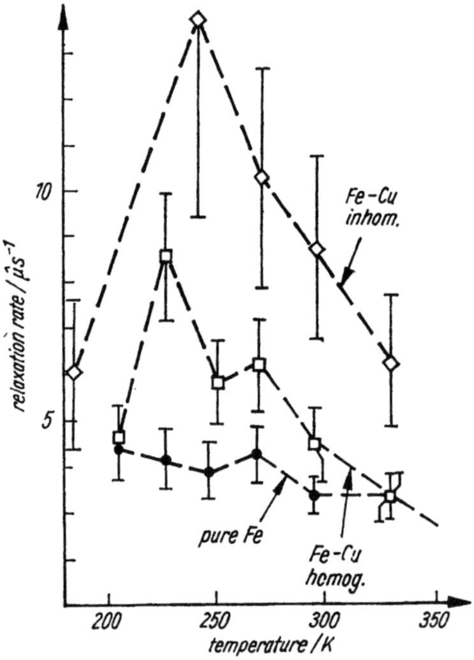

Fig. 1. Relaxation rates Λ for iron, homogeneous and precipitate-containing Fe-0.8 wt% Cu

i) comparing first the Fe–Cu samples with the pure iron solvent we see that the relaxation rates for the alloy, even in the homogeneous supersaturated phase, are below 300 K much higher than Λ(Fe), the difference being the largest between 200–250 K. The value $\Lambda = 3\ \mu s^{-1}$ at $T = 300$ K for our iron sample is somewhat higher than the reported $\sim 1\ \mu s^{-1}$ (GRAF et al.; DENISON et al.), indicating the presence of more impurities;
ii) comparison of the two alloy phases shows that Λ for the inhomogeneous phase is higher all over the measured temperature range, by a factor ~ 2, than that for the homogeneous solution;
iii) for both Fe–Cu specimens Λ has a maximum near $T_m = 240$ K, with values $8\ \mu s^{-1}$ and $14\ \mu s^{-1}$ for the homogeneous and inhomogeneous phases, respectively.

This indicates a diffusion-limited trapping mechanism: below T_m most of the implanted muons decay before attaining a trap, above T_m detrapping becomes more and more effective.

The signals for the steel (samples d) were strongly damped, corresponding to the large variety of possible traps in this material, and clearly oscillating signals (at least two periods) were seen only at $T = 330$ K. Here Λ for the unirradiated sample is about $15\ \mu s^{-1}$, whereas for the irradiated sample even larger, $18\ \mu s^{-1}$. This indicates an increased trapping rate due to the irradiation-induced defects, but a quantitative analysis requires further experiments in an extended, higher temperature range, where motional narrowing makes the relaxation time somewhat longer.

In conclusion, the presence of substitutional copper atoms (or eventually small agglomerates) in the homogenized sample leads already to a high trapping rate near 240 K, but precipitation of the copper (with particle radius 70–80 Å, as indicated by previous SANS data) increases Λ further, by a factor of ~ 2. This higher trapping rate is probably due to the large strain at the particle-solvent border regions, and if so, the smaller, radiation induced precipitates are expected to be even more effective muon traps.

Further experiments with irradiated samples are under progress.

References

BEAVEN, P. A., FRISIUS, F., KAMPMANN, R., WAGNER, R.: Atomic Transport and Defects in Metals by Neutron Scattering, ed. by C. JANOT, W. PETRY, D. RICHTER and T. SPRINGER, Berlin 1986, p. 228

DENISON, A. B., GRAF, H., KUENDIG, W., MEIER, P. F.: Helv. Phys. Acta **52**, 460 (1979)

FRISIUS, F., KAMPMANN, R., BEAVEN, P. A., WAGNER, R.: Dimensional Stability and Mechanical Behaviour of Irradiated Metals and Alloys, London 1983, p. 171

FRISIUS, F., NARAGHI, M.: Atomkernenergie **29**, 139 (1977)

GHAZI-WAKILI, K., ZIMMERMANN, U., BRUNNER, J., WAEBER, W. B., HEINRICH, F.: this conference (PSD 87, Wernigerode)

GRAF, H., KUENDIG, W., PATTERSON, B. D., REICHART, W., ROGGWILER, P., CAMANI, M., GYGAX, F. N., RUEGG, W., SCHENCK, A., SCHILLING, G.: Helv. Phys. Acta **49**, 730 (1976)

HIGHTON, J. P.: Radiation Embrittlement and Surveillance of Nuclear Reactor Pressure Vessels, ed. L. E. STEELE, ASTM STP 819, 1983, p. 117

LITTLE, E. A.: Effects of Irradiation on Materials: Twelfth Conference, ASTM STP 870, Met. All., London 1985, p. 1009

LUCAS, G. E., ODETTE, G. R.: "Proc. Second Intern. Symp. on Environmental Degradation of Materials on Nuclear Power Systems — Water Reactors", ANS 1986, p. 345

MOESLANG, A., GRAF, H., BALZER, G., RECKNAGEL, R., WEIDINGER, A., WICKART, Th., GRYNSZPAN, R. I.: Phys. Rev. **B27**, 2674 (1983)

Schenck, A.: Muon Spin Rotation Spectroscopy, Bristol 1985
Schwahn, D., Pachur, D.: 4th Int. Conf. on Nondestructive Testing, Lindau 25–27 May 1981, p. 629
Solt, G., Zimmermann, U., Waeber, W. B., Mercier, O., Frisius, F.: Int. Conf. on Structural Mechanics in Reactor Technology, Lausanne 1987, to appear
Steele, L. E.: Nuclear Safety **17**, 327 (1976)
Stronach, A., Squire, K. R., Arrott, A. S., Patterson, B. D., Heinrich, B., Lankford, W. E., Fiory, A. T., Kossler, W. J., Singh, J. I.: J. Magnet and Magn. Mater. **25**, 187 (1981)
Weidinger, A.: Proc. Yamada Conf. VII, Muon Spin Rotation 1983, Hyperfine Interactions **17–19**, 153 (1984)

(Received April 23, 1987)

Authors' addresses:

G. Solt, W. B. Waeber
Swiss Federal Institute for Reactor Research
CH-5303 Würenlingen, Switzerland

A. Schenck, F. N. Gygax, S. Barth
Institute for Medium Energy Research, ETH c/o SIN
CH-5234 Villingen, Switzerland

F. Frisius
Research Centre GKSS
D-2054 Geesthacht, FRG

Continued from cover page 4

Book Reviews

Crystal Research and Technology
Volume 22 1987 Number 11

Contents

Continued on cover page 3

Crystal Research and Technology is indexed in Current Contents/PC & ES.